Learning Guide to Physics for Scientists and Engineers

Learning Guide to Physics for Scientists and Engineers

by Fishbane, Gasiorowicz, and Thornton

Princeton University Department of Physics

Prentice Hall, Englewood Cliffs, New Jersey 07632

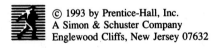 © 1993 by Prentice-Hall, Inc.
A Simon & Schuster Company
Englewood Cliffs, New Jersey 07632

All rights reserved. No part of this book may be
reproduced, in any form or by any means,
without permission in writing from the publisher.

Printed in the United States of America

10 9 8 7 6 5 4 3 2 1

ISBN 0-13-667247-7

Prentice-Hall International (UK) Limited, *London*
Prentice-Hall of Australia Pty. Limited, *Sydney*
Prentice-Hall Canada Inc., *Toronto*
Prentice-Hall Hispanoamericana, S.A., *Mexico*
Prentice-Hall of India Private Limited, *New Delhi*
Prentice-Hall of Japan, Inc., *Tokyo*
Simon & Schuster Asia Pte. Ltd., *Singapore*
Editora Prentice-Hall do Brasil, Ltda., *Rio de Janeiro*

Contents

Learning Guide	Text Chapter	Page
1	1, 2, 3	1
2	4, 5	13
3	6, 7	23
4	8	33
5	9, 10	43
6	11	53
7	12	63
8	13	69
9	14, 15	81
10	16	89
11	17, 18, 19	95
12	20, 21	101
13	22, 23, 24	111
14	25, 26	119
15	27, 28	129

Learning Guide	Text Chapter	Page
16	29, 30	137
17	31, 32, 33	147
18	34	157
19	35	179
20	36, 37	195
21	38, 39	203
22	41	209

Preface

These Learning Guides consist of a set of problems, Helping Questions, and keyed answers. The notation and the order in which the material is covered are matched to *Physics for Scientists and Engineers* by Fishbane, Gasiorowicz, and Thornton so that the text and Learning Guides complement each other. Unlike step-by-step "programmed learning" materials, these Learning Guides pose complete problems for students to solve. Help, which is given only when needed, comes in the form of increasingly detailed hints and references to the text.

The format is modeled after the sort of "one-on-one" interaction that occurs when a student attends office hours or obtains the help of an experienced tutor. Initially, a problem involving several steps is posed, and the student sets out toward the solution. If the student encounters difficulty along the way, the tutor often asks a "helping question," whose answer is needed to solve the original problem. The Learning Guides provide this same sort of guidance in written form. Since students needn't consult the Helping Questions unless they are having trouble, they can proceed quickly through the material that comes easily to them.

Each Learning Guide starts with a set of problems, which are often divided into parts. The answers to each part can be found by using the boldface **Key #**, which refers to a list of answers provided at the end of that Learning Guide. The list of Helping Questions also follows the problems. The answers to the Helping Questions are drawn from the same list as that for the problems.

Many of the Learning Guides come with supplementary notes, which complement the text by offering a different (and sometimes less formal) view of the material. In some cases, the notes go beyond the material covered in the text, for those who want extra topics or greater depth. Such material has been clearly marked.

First suggested by the late Professor Gerard K. O'Neill, the Learning Guides have been an integral part of the teaching of freshman physics at Princeton for more than two decades. Many faculty members, students, and staff of the physics department at Princeton have contributed to their development. Professor Daniel Marlow took on the job of editing them for publication. The department recognizes major contributions by dedicating this work to the memories of Professors Thomas R. Carver and Gerard K. O'Neill, who did so much to develop and improve physics instruction at Princeton.

—D. M.

learning guide 1

One-Dimensional Motion, Vectors, and Motion in a Plane

Suggested Reading: Fishbane, Gasiorowicz, & Thornton, Chapters 1, 2, and 3

PROBLEM I

A man is rowing a rowboat on a river whose current flows with speed v. The man can row at a speed V with respect to the water. Since $V > v$ he can row upstream as well as downstream. He decides to row a distance d (relative to the shoreline) upstream and then turn around and row downstream the same distance d to his starting point.

1. How long does it take the man to complete the round trip? If your answer doesn't check even after a good effort, look at Helping Questions 1 and 2.
       Key 16
2. Does the trip take longer when $v = 0$ or when $v > 0$? Why?   **Key 20**

PROBLEM II

A stone is dropped from the roof of an 80-m-high building. Its displacement, y, measured from the point of release is given by $y(t) = 5t^2$, where y is in meters and the time t is in seconds. In this problem you will calculate the stone's acceleration in two ways: by averaging over short time intervals, and by using calculus.

1. What is the displacement of the stone at $t = 0, 1, 2, 3$, and 4 s? What are the average velocities in the intervals 0–1 s, 1–2 s, 2–3 s, and 3–4 s? For small intervals, the instantaneous velocity at the center of an interval is approximately equal to the average velocity over the interval. Assuming for sake of argument that these two quantities are equal, what is the average acceleration in the time intervals 0.5–1.5 s, 1.5–2.5 s, and 2.5–3.5 s? If you're having trouble, use Helping Question 3. **Key 40**
2. Use calculus (i.e., take a derivative) to find the instantaneous velocity as a function of time. Find the instantaneous acceleration as a function of time. If you are stuck, see Helping Question 4. **Key 35**
3. If the physical situation had been different so that the displacement function was $y(t) = 5t^4$ instead of $y(t) = 5t^2$, which method would have been more accurate for calculating the instantaneous acceleration? **Key 8**

PROBLEM III: VECTOR WARM-UPS

1. Let $\mathbf{A} = 3\mathbf{i} + 4\mathbf{j}$ and let $\mathbf{B} = 5\mathbf{i} + 12\mathbf{j}$. What is the magnitude of $\mathbf{A}$? What is the magnitude of $\mathbf{B}$? What is $\mathbf{A} + \mathbf{B}$? What is the angle between $\mathbf{A}$ and $\mathbf{B}$? If you're having trouble, look at Fishbane, Gasiorowicz, & Thornton Section 1-6. **Key 31**
2. Now let $\mathbf{A} = 3\mathbf{i} - 4\mathbf{j}$ and let $\mathbf{B} = -5\mathbf{i} - 12\mathbf{j}$. What is the magnitude of $\mathbf{A}$? What is the magnitude of $\mathbf{B}$? What is $\mathbf{A} + \mathbf{B}$? What is the angle between $\mathbf{A}$ and $\mathbf{B}$? **Key 14**
3. Show the region in the xy-plane that contains the end points of all possible vectors $\mathbf{A} + \mathbf{B}$ where $\mathbf{A}$ is a vector of magnitude 5 and $\mathbf{B}$ is a vector of magnitude 13. **Key 1**

PROBLEM IV

An amateur football player punts a football straight up at $v_0 = 15$ m/s. Assume that the acceleration due to gravity is $g = 10$ m/s^2.

1. How long is the football in the air? If you need help getting started, use Helping Questions 5 and 6. **Key 45**
2. How high does the football go? Helping Question 7 will put you on the right track. **Key 9**

Next, a professional football player punts a football straight up. His football leaves his foot at twice the speed of the amateur player's football; i.e., with $v_0 = 30$ m/s.

3. Does the ball stay in the air twice as long? **Key 18**
4. Does it go twice as high? **Key 3**

Learning Guide 1   One-Dimensional Motion, Vectors, and Motion in a Plane

PROBLEM V

A grazing antelope first notices a lion attacking when the lion is 12.5 m away and moving toward the antelope at a speed of 5 m/s. The antelope begins to accelerate away from the lion at 3 m/s² and the lion simultaneously begins to accelerate at 2 m/s².

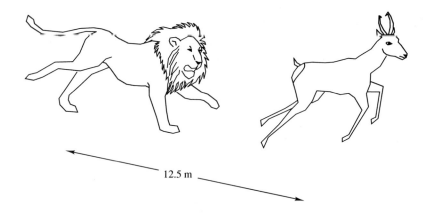

1. How long does the antelope's flight last? **Key 26**
2. How far has the antelope traveled when the lion catches up with it? **Key 5**

If you need help see Helping Question 8.

PROBLEM VI

An object dropped from a stationary helicopter falls straight down toward the earth. At the beginning of its flight, its acceleration is g, the acceleration due to gravity. As its speed increases, it meets with increasing amounts of *air resistance*. To a reasonable approximation, its motion satisfies $a(t) = g - kv(t)$ where $a(t)$ and $v(t)$ are the instantaneous acceleration and velocity and k is a positive constant that depends on the shape and surface roughness of the object.

1. *Without* solving an equation for $v(t)$ use physical reasoning to sketch a graph of speed versus time. You should discover the **terminal velocity** phenomenon: the object never accelerates past a particular speed, called the *terminal velocity*. What is the terminal velocity for the object in terms of g and k? Turn to Helping Questions 9 and 10 if you need help. **Key 7**

PROBLEM VII

In Problem II, you were given $x(t)$ and asked to find $v(t)$ and $a(t)$. Suppose you were given $a(t)$ and asked to find $v(t)$ and then $x(t)$. In this problem, you will learn a geometrical solution to this question. You know that

$$v(t) = v_0 + at$$

for motion with constant acceleration. Referring to the graph of $a(t)$ in the figure, note that

$$v(t) = v_0 + \text{area under the curve between } O \text{ and } t, \tag{1}$$

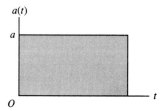

since the shaded area is precisely at. In fact, it turns out that equation (1) is generally true, even if the acceleration is *not* constant. In such cases, the equation $v = v_0 + at$ no longer makes sense (since a is no longer a single number).

As an example, consider the next sketch, where $a(t)$ varies with time. The area of the shaded region is about 6 m/s, and so the velocity at $t = 3$ s is $v_0 + 6$ m/s.

In the same way, the equation of motion, $x(t) = x_0 + vt$, not only applies for motion at constant v, but generalizes to

$$x(t) = x_0 + \text{area under the } v(t) \text{ curve between 0 and } t$$

no matter how the velocity changes with time.

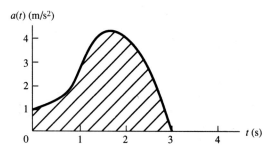

For both the acceleration and the velocity curves, a slight complication enters if the curves go beneath the horizontal axis—i.e., if the acceleration or velocity goes negative. Then "area under the curve" must be replaced by "area above the horizontal axis *minus* the area below the horizontal axis," as shown in the sketch. You should convince yourself that this extension is physically reasonable. In the example shown, the position at $t = 3$ s is about $x(3) = x_0 + 1$ m. (Think about where the particle is at the top of the first "+" bump.)

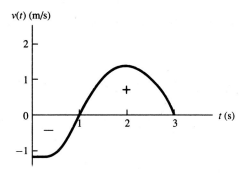

For practice, consider a particle moving with acceleration given by the graph below. Take $v_0 = 0$ and $x_0 = 0$. Graph as accurately as you can, putting numbers on the axes:

1. $v(t)$ **Key 2**
2. $x(t)$ **Key 29**

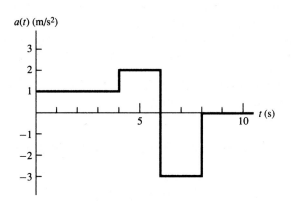

PROBLEM VIII

A man can row a boat 8 km/h in still water. He is on one bank of an 8-km-wide river that has a current of 4 km/h.

1. What is the smallest amount of time he needs to get to the other bank, assuming that it's acceptable to land anywhere on the other shore? If you're confused, see Helping Question 11.   **Key 34**
2. If the man wants to get to the point exactly across from where he starts, what angle should his boat make with respect to a line running straight across the river? Helping Question 12 will help you with this part. **Key 12**
3. Take the point of view of an observer on the shore. In your reference frame, the man's velocity vector is the sum of a vector of length 8 km/h (his rowing speed in still water) and a vector of length 4 km/h (the speed of the river's current). In your reference frame, what is the man's velocity in part (1)? In part (2)?   **Key 19**

If you need some hints, turn to Helping Questions 13 and 14.

PROBLEM IX

A cannon fires a projectile at an angle θ with respect to the horizontal. The speed of the projectile as it leaves the cannon's barrel is v_0. Find an expression that gives the horizontal range of the shell, R, in terms of θ, v_0, and g. See Helping Questions 15 and 16. (This is the **range formula**, which is also derived in Fishbane, Gasiorowicz, & Thornton, but don't look there unless you're really stuck.)   **Key 33**

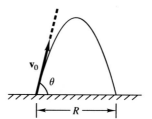

PROBLEM X

A man standing on a cliff of height h has a baseball that he can throw with speed v. He wants to throw the baseball as far away from the cliff as he can. In terms of the variables indicated on the diagram, he wants to choose θ to maximize d. For this problem you may neglect the height of the man, air resistance, and the bouncing or rolling of the ball. Before you start, you might want to guess at θ_{max} for $h = 0$ and h very high. **Note:** The range formula from Problem IX does *not* apply. It is valid only in cases where the initial and final heights are the same.

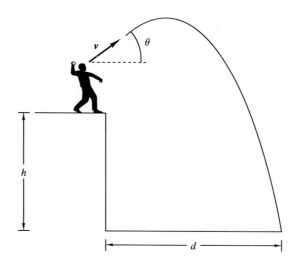

1. Express d in terms of v, θ, h, and g, the acceleration due to gravity. Use Helping Questions 17 and 18.   **Key 15**
2. Set $h = 0$ in your answer for part (1). What is θ_{max}? Does θ_{max} depend on v? For a hint, see Helping Question 19.   **Key 11**

3. Now suppose that h is "infinitely large" so that terms not containing h can be ignored. What is $\theta_{\max}$? Does $\theta_{\max}$ depend on v? Stumped? See Helping Question 20. **Key 6**

4. Finally, consider the case where h is finite and positive. Between what two values of θ is $\theta_{\max}$? Use your physical intuition to decide whether $\theta_{\max}$ depends on v. Use Helping Question 21 for the last question. **Key 42**

HELPING QUESTIONS

1. What is the man's speed relative to the shoreline when he rows upstream? When he rows downstream? **Key 39**
2. How long does it take him to row upstream? Downstream? **Key 10**
3. What is the definition of average velocity? Average acceleration? **Key 24**
4. Define instantaneous velocity and instantaneous acceleration mathematically. **Key 32**
5. Can you think of a kinematic formula that relates what you know—the starting point $y_0 = 0$, the end point $y = 0$, v_0, and g—to what you want to find, which is the time of landing t? **Key 37**
6. What are the two roots of the equation $v_0 t - \frac{1}{2} g t^2 = 0$? **Key 25**
7. You know the time at which the ball hits the ground. Can you say right away the time at which the ball is at the high point of the path? If you don't see the intuitive answer, again try to find the right equation. The key step is to think of what quantity has a special behavior at the top. **Key 30**
8. In what is true about the two animals' positions at the time of interest? **Key 46**
9. Is the velocity of the object increasing or decreasing? Is the acceleration of the object increasing or decreasing? What do the answers to these two questions mean in terms of the plot of $v(t)$? **Key 27**
10. Suppose the object was falling with terminal velocity. What would the acceleration be? **Key 38**
11. Does the speed of the river's current affect the time the man needs to cross? **Key 17**
12. In what direction must the velocity vector (rowing velocity plus current velocity) be pointing? Make a sketch, showing all three vectors and the angle you're after. **Key 28**
13. One way you can add vectors is to introduce a coordinate system and add the vectors by component. What would be a convenient coordinate system here? **Key 13**
14. If a vector **v** makes an angle θ with the x-axis, what is v_x in terms of $|\mathbf{v}|$ and θ? What is v_y? **Key 23**
15. What is $x(t)$? What is $y(t)$? If you need another hint, move on to Helping Question 16. **Key 36**
16. $y(t)$ has two zeros; what is their significance? **Key 4**
17. If you knew the time t the ball spends in the air, could you get d in terms of θ, v, and t? **Key 41**
18. Can you get the time of flight t from the vertical component of the motion? You'll have to use the quadratic formula. **Key 21**
19. A trigonometric identity says that $2 \sin\theta \cos\theta = \sin(2\theta)$. Can you

get the maximum by thinking about the graph of $\sin(2\theta)$? Alternatively, you can use calculus, setting the derivative of d with respect to θ equal to zero. **Key 44**

20. What is the expression for d after the terms not containing h have been ignored? **Key 22**

21. Think of a little boy and a professional baseball player on a 5-m cliff. The baseball player has a much higher v, of course. To the little boy, is the cliff extremely high or negligible? What about to the baseball player? **Key 43**

Notes: Dimensions

You should read Sections 1-2 and 1-3 of Fishbane, Gasiorowicz, & Thornton about units and dimensions very carefully. As this course progresses, you will appreciate more and more their statement that *the dimensions on one side of an equation must be the same as those on the other side*. In fact, in any legitimate physical equation the dimensions of *all the terms* must be the same. Perhaps more to the point, a look at the dimensions of a solution to a problem often provides a quick "sanity check" on its veracity. If, for example, you are asked to calculate a velocity or a speed, your answer had better have units of meters per second. Any other dimensions indicate that something has gone awry. Old pros employ this trick to check their work as standard practice.

One can even check dimensions in equations that contain derivatives. The dimensions of a derivative are identical to the dimensions of the corresponding fractions. Thus, for example, dx/dt has the same dimensions as x/t. Later on you will learn how to use integrals in physics, but from the point of view of dimensions an integral is just like a multiplication. You will also become familiar with algebraic operations between vectors—vector addition and two different types of vector multiplication. Dimensionally, these operations are the same as normal addition and multiplication.

The following section is optional reading ...

One can even use dimensional analysis to make "ballpark" estimates. To illustrate this, let's try to solve Problem IV about punted footballs using only dimensional arguments and intuition. Part (1) asks for the time t that the ball is in the air. Intuitively, the greater the initial speed v_0, the greater the time of flight t. Intuition also says that the greater gravity g is, the smaller t is. So a possible solution is:

$$t = \frac{v_0}{g} \qquad \left(\frac{[L]/[T]}{[L]/[T]^2} = \frac{1}{1/[T]} = [T] \right)$$

This equation checks dimensionally, as the computation in parentheses shows (in that computation, $[L]$ and $[T]$ denote quantities having dimensions of length and time, respectively). In fact, the exact solution is $t = 2v_0/g$.

This is about as close as dimensional analysis can get, since the "2" is dimensionless and is typical of the kind of accuracy one can expect using this approach.

Part (2) asks for the maximum height h of the ball. Again, intuition suggests a possible solution:

$$h = \frac{v_0}{g} \qquad \left(\frac{[L]/[T]}{[L]/[T]^2} = [T] \neq [L]\right).$$

But this equation does not check dimensionally, since $[L] \neq [T]$. If you think about it, the only simple way to fix it up is to write:

$$h = \frac{v_0^2}{g} \qquad \left(\frac{([L]/[T])^2}{[L]/[T]^2} = \frac{[L]^2/[T]^2}{[L]/[T]^2} = [L]\right),$$

which does check dimensionally. Again, it is off from the exact expression, $h = v_0^2/2g$, by a dimensionless factor of 2.

The real triumph of dimensional analysis in this example is the answers it gives for parts (3) and (4). Even though it is off by dimensionless factors, it still predicts correctly that if the initial speed is doubled, the time of flight is also doubled but the maximum height is multiplied by 4.

ANSWER KEY

1.
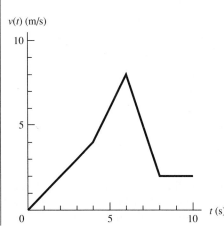

Inner circle has radius 8.
Outer circle has radius 18.

2.

3. No—it goes 4 times higher.
4. Launch time and landing time
5. 37.5 m
6. $\theta_{max} = 0°$, independent of v.
7. $v_t = g/k$

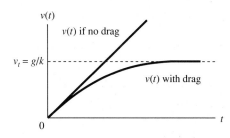

8. The calculus method of part (2) would have been better. No matter how small the time intervals were made in part (1), the answer would still be a little off.
9. 11.25 m
10. $d/(V - v)$ upstream; $d/(V + v)$ downstream

11. $\theta_{max} = 45°$, independent of v.

12. 30° upstream

13. Let the x-direction be downstream and the y-direction be across the river.

14. $|\mathbf{A}| = 5$, $|\mathbf{B}| = 13$, $\mathbf{A} + \mathbf{B} = -2\mathbf{i} - 16\mathbf{j}$, $\theta_{AB} = 59.5°$

15. $d = v\cos\theta \left(v\dfrac{\sin\theta}{g} + \sqrt{\dfrac{v^2 \sin^2\theta}{g^2} + \dfrac{2h}{g}} \right)$

16. $\dfrac{d}{V-v} + \dfrac{d}{V+v} = \dfrac{2dV}{V^2 - v^2}$

17. No

18. Yes

19. $\sqrt{80} \simeq 9$ km/h for part (1) and $\sqrt{48} \simeq 7$ km/h for part (2).

20. The trip takes longer when $v > 0$ because when $v > 0$ the man's average speed is less than V: he spends more than half the time at speed $V - v$ and less than half the time at speed $V + v$.

21. $t = \left(\dfrac{v \sin\theta}{g} + \dfrac{1}{g}\sqrt{v^2 \sin^2\theta + 2gh} \right)$

22. $d = v \cos\theta \sqrt{\dfrac{2h}{g}}$

23. $v_x = |\mathbf{v}| \cos\theta$; $v_y = |\mathbf{v}| \sin\theta$

24. $v_{av} = \dfrac{\Delta y}{\Delta t}$; $a_{av} = \dfrac{\Delta v}{\Delta t}$

25. The starting time $t = 0$ and the landing time $t = \dfrac{2v_0}{g}$.

26. 5 s

27. The speed is increasing but the acceleration is decreasing, so the slope of the speed curve will always be positive but will become less steep at larger t.

28. Straight across the river

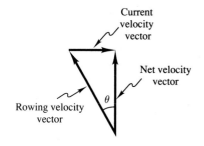

29.

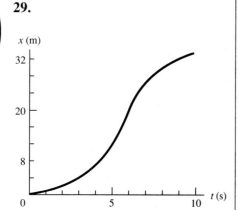

30. Intuitive method: It's at the top at the midpoint of the trip, or $t = 1.5$ s. Systematic method: The key point is that the ball stops instantaneously at the top, i.e., $v = 0$. Use $0 = v_0 - gt$ to solve for the time to get to the top.

31. $|\mathbf{A}| = 5$, $|\mathbf{B}| = 13$, $\mathbf{A} + \mathbf{B} = 8\mathbf{i} + 16\mathbf{j}$, $\theta_{AB} = 14.3°$

32. $v = \dfrac{dy}{dt}$; $a = \dfrac{dv}{dt}$

33. $R = \dfrac{2v_0^2}{g} \sin\theta \cos\theta = \dfrac{v_0^2}{g} \sin(2\theta)$

34. 1 h

35. $v = 10t$, $a = 10$, so the acceleration is constant at 10 m/s².

36. $x(t) = (v_0 \cos\theta)\, t$;
    $y(t) = (v_0 \sin\theta)\, t - \dfrac{1}{2} g t^2$

37. $y = y_0 + v_0 t + \tfrac{1}{2} a t^2$; $a = -g$

38. 0

39. $V - v$ upstream; $V + v$ downstream.

40. For $t = 0, 1, 2, 3, 4$ s:
    $y = 0, 5, 20, 45, 80$ m

$v_{av} = 5, 15, 25, 35$ m/s
$a_{av} = 10, 10, 10,$ m/s^2
41. $d = (v \cos\theta) t$
42. θ_{max} is between 0° and 45° and depends on v.
43. High to the boy, negligible to the player
44. $\sin(2\theta)$ has a maximum at $\theta = 45°$.
45. 3 s
46. They are equal; i.e., at the time of interest $x_{lion}(t) = x_{antelope}(t)$.

learning guide 2

Newton's Laws

Suggested Reading: Fishbane, Gasiorowicz, & Thornton, Chapters 4 and 5

PROBLEM I

A 100-kg man is standing on a bathroom scale in an elevator. Take the acceleration g due to gravity to be $g = 10$ m/s^2.

1. Suppose that the elevator is moving at a constant speed of 2 m/s. What is the reading on the scale? (The scale is calibrated in kilograms, not in newtons.) **Key 36**
2. Suppose that the elevator is accelerating up at the constant rate of 2 m/s^2. What does the scale read? **Key 1**

PROBLEM II

Two blocks, in contact on a frictionless table, are pushed by a horizontal force applied to one block, as shown in the figure.

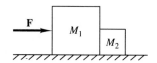

1. What is the force of contact between the two blocks in terms of F, M_1, and M_2? Answering Helping Question 1 will get you started.   **Key 19**

In a second situation, a force of equal magnitude but opposite direction is applied to the other block.

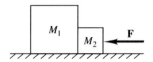

2. What is the force of contact in this situation? If $M_1 \neq M_2$, is the force the same as in part (1)?   **Key 33**

PROBLEM III

A block of mass M lies on a horizontal surface. A force $\mathbf{F}$ acts on the block at an angle θ from the horizontal. The coefficient of static friction is μ_s.

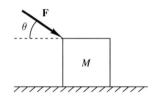

1. What is the minimum force necessary to start the block moving as a function of the angle θ? Use Helping Questions 2, 3, and 4 if necessary.   **Key 20**
2. From your answer to part (1), can you find a critical angle θ_c such that for angles θ greater than θ_c even huge forces won't move the block? If $\mu_s = 0.4$, what is θ_c? Turn to Helping Question 5 if you're confused.   **Key 8**
3. Think physically about why there is a critical angle past which the block won't slide. What's increasing with θ and what's decreasing as θ increases?   **Key 34**

PROBLEM IV

In the apparatus shown in the sketch, the string is massless and does not stretch, and the pulley is massless and has frictionless bearings. The coefficient of kinetic friction between block 1 and the table is μ_k.

Learning Guide 2   Newton's Laws

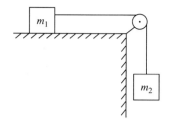

1. Assuming that m_2 is large enough that the blocks move, find the acceleration of the blocks. Use Helping Questions 6, 7, and 8, if you need to. **Key 39**

PROBLEM V

A woman driving her car 60 mi/h down a country road sees that a huge tree has fallen across the road some distance in front of her. She immediately slams on her brakes and skids to a stop. Fortunately the pavement is dry and the coefficient of kinetic friction between her tires and the road is $\mu_k = 0.5$.

1. How many feet does it take the car to stop? (60 mi/h equals 88 ft/s.) Use $g = 32$ ft/s². If you're really stuck, use Helping Questions 9 and 10 sparingly. **Key 42**

PROBLEM VI

Two blocks of mass $m_1 = 10$ kg and $m_2 = 2$ kg slide along a horizontal surface. An external force F applied to m_1 from the left provides enough acceleration to keep m_2 from sliding down the face of m_1. The coefficient of *kinetic* friction between m_1 and the horizontal surface is $\mu_k = 0.3$ and the coefficient of *static* friction between m_1 and m_2 is $\mu_s = 0.5$. Take g to be 10 m/s².

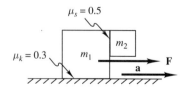

1. Assuming that m_2 does not slip, find a, the acceleration of the two blocks when $F = 350$ N. If need be, see Helping Question 11. **Key 40**
2. For $F = 350$ N, what are the components of $\mathbf{F}_{21}$, the force exerted on m_2 by m_1? See Helping Question 12. **Key 14**
3. What is the smallest value of F required to keep m_2 from slipping? See Helping Question 13. **Key 18**

PROBLEM VII

An observer in an elevator can't distinguish between his own acceleration and the earth's gravity. In the same way, the acceleration caused by uniform circular motion feels like gravity. On the earth's surface the acceleration due to gravity is, to three significant figures, 981 cm/s². But if an experimenter at the equator measured the acceleration of objects falling in a vacuum, he would get a number less than this because of the earth's rotation.

1. To three significant figures, what acceleration would be measured? (The equatorial radius of the earth is about 6000 km.) See Helping Question 14, if you need it. **Key 7**
2. Why were you able to use such a crude approximation for the earth's radius when you wanted three-figure accuracy? **Key 15**

PROBLEM VIII

A stunt man rides a bobsled through a loop-the-loop track as shown in the diagram. The loop has radius R.

1. Find the minimum speed v_t at the top of the track that ensures that the sled will stay on the track. See Helping Questions 15 and 16. **Key 4**

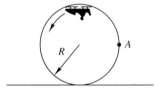

2. On a second pass through the loop, the sled has speed v_A when it reaches point A on the loop. Find the magnitude and direction of the *net* force acting on the sled at point A. For help, see Helping Questions 17 and 18. **Key 27**

PROBLEM IX

An engineer designing the bank on a curved road has to take friction into account. There are a lot of variables, and so she might use a graph such as the one below to organize her thoughts. For this graph the radius of the curve R and the coefficient of friction μ_s are fixed. The solid line is the "perfect" bank: the car goes around the curve and the road exerts *no sideways frictional force* on the car. The dashed lines denote the steepest possible angle and the gentlest possible bank angles θ that allow the car to negotiate the curve without skidding sideways. In parts (1) to (3) of this problem, you will get equations for these three curves.

Learning Guide 2   Newton's Laws

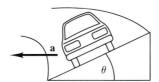

1. For what speed v is the angle θ a perfect bank? Express v in terms of θ, g, and R. Helping Questions 19 and 20 will give you the method of attack, but by now you should know it. **Key 13**

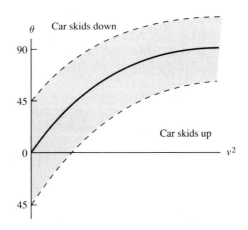

2. At what speed v will the car just begin to slide *up* a bank of angle θ? Express v in terms of θ, g, R, and μ_s. **Key 3**
3. At what speed v will the car just begin to slide *down* a bank of angle θ? Write v in terms of θ, g, R, and μ_s. **Key 22**
4. The adjacent diagram shows a curve banked *past vertical* so that $\theta > 90°$. Looking at your answer to part (3), if $\mu_s > 0$, is there a speed great enough that the car stays on the road? For a hint, see Helping Question 21. **Key 30**

5. The graph above has a certain *symmetry*: if for a speed v the perfect bank is θ and the car will just begin to slide down at $\theta + \Delta\theta$, then, the way the graph is drawn, it will just begin to skid up at $\theta - \Delta\theta$. Is this correct? The equations from parts (2) and (3) are a little messy, so try using physical reasoning. If need be, see Helping Question 22. **Key 32**

HELPING QUESTIONS

1. What is the acceleration of the two blocks? The force on the right block? **Key 16**

2. Draw a force (free-body) diagram and label all the forces. What is the frictional force when the applied force F is at its maximum value before slipping? **Key 26**

3. What is the acceleration a when the applied force F is at its maximum value—i.e., just before the block starts slipping? **Key 6**

4. What is the horizontal component of Newton's second law $\mathbf{F} = m\mathbf{a}$? The vertical component? **Key 37**

5. As θ increases from zero, the minimum force F required also increases. For what value of θ is the minimum force required infinite? **Key 29**

6. Draw a force (free-body) diagram for each block and apply $\mathbf{F} = m\mathbf{a}$ to each block. Let T_1 be the force of the string on block 1 and T_2 its force on block 2. Let a_1 be the acceleration of block 1 and a_2 the acceleration of block 2. **Key 5**

7. Is there a relation between T_1 and T_2? **Key 31**

8. Is there a relation between a_1 and a_2? **Key 43**

9. Draw a force (free-body) diagram and find the resultant force on the car. **Key 41**

10. Can you reduce the problem to a kinematics problem with constant acceleration? **Key 12**

11. If m_1 doesn't slip is there any difference between this situation and the case of a single block with mass $m_1 + m_2$? **Key 2**

12. What *other* force or forces act on m_2? Given these, Newton's second law, and the acceleration from part (1), one can deduce the components of $\mathbf{F}_{21}$. **Key 23**

13. What is the physical significance of each component of $\mathbf{F}_{21}$? **Key 10**

14. You know that the earth rotates once every 24 h. What is the acceleration v^2/R of a point on the equator? **Key 9**

15. What forces act on the sled? What are the magnitude and direction of its acceleration? **Key 17**

16. What is the normal force just at the point when the sled is about to fall? **Key 11**

17. What is the force in the vertical direction? **Key 21**

18. What equation expresses Newton's second law in the horizontal direction? **Key 24**

19. Draw a force (free-body) diagram for the car and identify the acceleration vector. **Key 28**

20. What is the horizontal component of $\mathbf{F} = m\mathbf{a}$? The vertical component? **Key 38**

21. When does the denominator of the answer to part (3) equal zero? **Key 25**

22. An observer in the car can't distinguish between gravity and his own acceleration. Imagine that he decides that he is not accelerating but rather that the force from gravity is greater than usual and angled strangely. How would he draw the force diagram for the car when the bank is perfect? **Key 35**

Notes: More about Dimensional Analysis

Mentally Varying Variables

As you can see by now, a big part of learning physics is learning how to manipulate and understand equations. The first and most important thing to learn is to manipulate the equations while they still are in variable form and only plug in the numbers at the end. Indeed, the ability to solve problems using algebra allows us to extend our mental "grasp" in a very profound way. Learn to take advantage of algebra, one of man's great achievements!

This Learning Guide has tried to force you to acquire this habit by not always giving you numbers. There are at least two advantages of not plugging in. The first relates back to dimensional analysis. If you choose to plug in the numbers from the start of a problem, you will quickly lose track of the dimensions unless you are *extremely* careful (in which case you'd be taking our advice!).

Indeed, you don't need to wait until the end of a problem before making dimensional checks. Certain quantities are always dimensionless. For example, whenever you see $\sin x$, $\cos x$, or any trigonometric function, *its argument (x) must be dimensionless*. This is because if you changed units from, say SI to the British system, the angle must not change (angles are always the ratio of two like-dimensioned quantities). Also, the *value of any trigonometric function is dimensionless* (such quantities are even more obviously the ratio of like dimensioned quantities).

Another advantage has to do with *mentally varying the variables in any equation*. This not only provides another type of "sanity check" but is an excellent way to understand what an equation really means. As an example, let's try to fully understand the solution to the problem of the two blocks and the pulley,

$$a = g \left(\frac{m_2 - \mu_k m_1}{m_2 + m_1} \right).$$

First, the equation is dimensionally correct since a and g are both accelerations and the factor in parentheses is dimensionless (coefficients of friction, being the ratio of two forces, are dimensionless). Now imagine that m_1 was very small or even zero. Then

$$a = g \frac{m_2}{m_2} = g,$$

which you recognize as the acceleration of m_2 freely falling. Imagine instead that m_1 was very big, so big that m_2 was negligible. Then

$$a = g \left(\frac{-\mu_k m_1}{m_1} \right) = -g\mu_k.$$

This equation is clearly *incorrect* because the masses certainly don't accelerate backward: they just sit there if m_1 is too massive. If you think about it, you have just discovered the critical condition for acceleration: m_2 must be greater than $\mu_k m_1$. For practice, further your understanding of this equation by seeing what happens to a when you (1) vary g, (2) vary μ_k, or (3) double m_1 and m_2 simultaneously. Your answers should match your physical intuition. Of course, if you just started with something like $a = 3.7$ m/s^2, you couldn't do any of this.

This method of mentally varying variables to extremes is also a good way to get a feeling for a problem before you actually begin writing equations. For the problem of pushing the block hard enough to overcome static friction, if you let θ be 90° and push straight down then, of course, the block doesn't move! So, it's not too surprising that there is a critical angle near 90°. For the problem of two blocks in contact on a table, take one of them to be huge—say, your physics textbook—and one of them to be tiny—perhaps a little ball of clay. Of course, the forces of contact are different—if you push your text with a little ball of clay, the ball immediately squishes, but if you push the ball with your text the ball just rolls away!

ANSWER KEY

1. 120 kg
2. No
3. $$v = \sqrt{gR\left(\frac{\sin\theta + \mu_s \cos\theta}{\cos\theta - \mu_s \sin\theta}\right)}$$
4. $v_t = \sqrt{gR}$
5.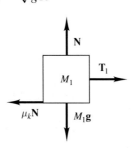

so $N = M_1 g$
and $T_1 - \mu_k N = M_1 a_1$.

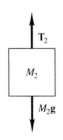

so $M_2 g - T_2 = M_2 a_2$.

6. 0
7. 978 cm/s^2
8. $\theta_c = \cot^{-1} \mu_s = 68°$
9. 3 cm/s^2
10. F_{21_x} is the normal force; F_{21_y} is the force of static friction.
11. 0 (N can't be negative)
12. Yes, the constant acceleration is $\mu_k g = -16$ ft/s^2.
13. $$v = \sqrt{gR \tan\theta}$$
14. $F_{21_x} = m_2 a = 52.3$ N; $F_{21_y} = m_2 g = 20$ N
15. The correction due to the earth's rotation is small, contributing only to the third decimal place.
16. Both blocks have acceleration $F/(m_1 + m_2)$. The force on the right block is $m_2 F/(m_1 + m_2)$.
17. Gravity and the normal force, both straight down. $a = v_t^2/R$ inward (i.e., straight down).
18. $$F = \left(\frac{1}{\mu_s} + \mu_k\right) \times (m_1 + m_2) g = 276 \text{ N}$$

19. Force of contact is
$$F_{contact} = F\left(\frac{M_2}{M_1 + M_2}\right).$$

20. Minimum force for movement is
$$F_{min} = \frac{\mu_s M g}{\cos\theta - \mu_s \sin\theta}.$$

21. mg (down)
22. $$v = \sqrt{gR\left(\frac{\sin\theta - \mu_s \cos\theta}{\cos\theta + \mu_s \sin\theta}\right)}$$

23. $m_2 g$ (acting down)
24. $N = mv^2/R$, where N is radially inward
25. When $\cot\theta = -\mu_s$
26.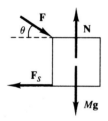

At the maximum value of F before slipping, $F_s = \mu_s N$.

27. $$|\mathbf{N} + m\mathbf{g}| = m\sqrt{\frac{v_A^4}{R^2} + g^2}$$

$\theta = \tan^{-1}\left(\frac{gR}{v_A^2}\right)$ below the horizontal.

28.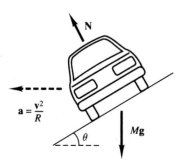

29. When the denominator is zero

30. There is a speed high enough if $\cot\theta > -\mu_s$. Otherwise the car will slide down or fall off.
31. $T_1 = T_2$ (the pulley has no mass to accelerate).
32. Yes
33. $$F_{contact} = F\left(\frac{M_1}{M_1 + M_2}\right),$$

which is different from part (1).

34. Normal force, and thus the frictional force, increases with θ. The horizontal component of $\mathbf{F}$ decreases as θ increases. Both effects discourage sliding.

35.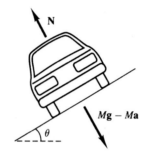

36. 100 kg
37. Horizontal component:
$$F\cos\theta - \mu_s N = 0$$

Vertical component:
$$mg + F\sin\theta - N = 0$$

38. Horizontal: $N\sin\theta = \dfrac{mv^2}{R}$

Vertical: $N\cos\theta = mg$

39. $$a = g\left(\frac{m_2 - \mu_k m_1}{m_1 + m_2}\right)$$

40. $$a = \frac{F}{m_1 + m_2} - \mu_k g = 26.2 \text{ m/s}^2$$

41. 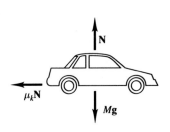 so resultant force is $\mu_k mg$, slowing the car down.

42. 242 ft

43. $a_1 = a_2$ (the blocks are tied together and the string doesn't stretch).

learning guide 3

Work and Energy

Reading Assignment: Fishbane, Gasiorowicz, & Thornton, Chapters 6 and 7

VECTOR WARM-UPS—THE SCALAR PRODUCT

1. What is **a** · **b**? If you don't remember what the definition of the scalar product is, go back to Fishbane, Gasiorowicz, & Thornton, Section 1-6.
Key 25

There is another method of calculating scalar products that is occasionally more convenient than the formula $\mathbf{a} \cdot \mathbf{b} = |\mathbf{a}||\mathbf{b}|\cos\theta$. You will discover this method in parts (2) and (3) and then use it in part (4).

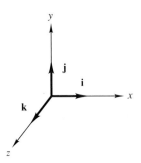

2. **i**, **j**, and **k** are the unit vectors in the coordinate system drawn in the diagram. Fill in the multiplication table:

•	i	j	k
i			
j			
k			

    If you need help, turn to Helping Question 1. **Key 6**

3. In a certain fixed coordinate system $\mathbf{a} = a_x\mathbf{i} + a_y\mathbf{j} + a_z\mathbf{k}$ and $\mathbf{b} = b_x\mathbf{i} + b_y\mathbf{j} + b_z\mathbf{k}$. In terms of the components a_x, b_x, a_y, b_y, a_z, and b_z, what is $\mathbf{a} \cdot \mathbf{b}$? Stuck? Turn to Helping Question 2. **Key 18**

4. Suppose the vectors in part (1) are now given in terms of their components in a rectangular coordinate system. What is $\mathbf{a} \cdot \mathbf{b}$? **Key 32**

PROBLEM I

A stone is released from a height h at time $t = 0$ and falls straight down under the influence of gravity. As you know, its height y at time t is given by $y(t) = h - gt^2/2$.

1. How much work W has the force of gravity done on the stone from the time of its release to time t? If you need help, use Helping Question 3. **Key 17**

As you also know, the stone's speed v at time t is given by $v = -gt$.

2. From the definition of kinetic energy, what is the kinetic energy K_0 at the start, and what is the kinetic energy K at time t? **Key 31**
3. Verify the work energy theorem $W = \Delta K$ for this system. **Key 7**
4. From the definition of instantaneous power $P = dW/dt$, what is the power delivered by gravity to the stone at time t? From the formula $P = \mathbf{F} \cdot \mathbf{v}$, what is the power delivered by gravity to the stone at time t? Do your two answers agree? **Key 38**

Learning Guide 3   Work and Energy

PROBLEM II

A man wants to lift a block of weight $W = 1000$ lb a height $h = 5$ ft off the ground. He attempts to contrive a way to avoid doing 5000 ft·lb of work on the block. One idea he thinks up is to employ $2n$ pulleys, arranged as shown in the sketch. Assume that the tension is the same everywhere in the rope and that friction and the rope's deviation from the vertical can be neglected.

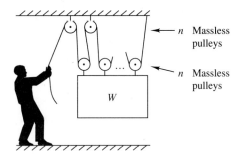

1. Using a force (free-body) diagram, determine the magnitude of the force F the man needs to exert to lift the block in terms of W and n. Assume that the initial acceleration of the block is negligible and that, once in motion, the block moves with constant speed. Use Helping Questions 4 and 5 if you need to. **Key 23**
2. Through what distance d does the man need to pull the rope in order to lift the block h ft? Use Helping Question 6, if necessary. **Key 5**
3. Does the contraption decrease the amount of work he has to do? **Key 8**

PROBLEM III

A mass m is attached to a massless spring (unstretched length l_0, and spring constant k) and moves in a circular path of radius R. Assume at first that there is no friction between the mass and the horizontal surface.

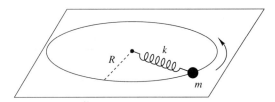

1. Find the ratio of the potential energy in the spring to the kinetic energy of the mass in terms of known quantities—i.e., m, l_0, k, R, g. After a good try, use Helping Question 7. **Key 1**
2. Can the spring's potential energy equal or exceed the mass's kinetic energy? **Key 26**
3. Now, suppose a demon suddenly "turns on friction" between the mass and the surface. Find the distance d that the mass moves before it stops. Assume

that the coefficients of static and kinetic friction are both very small and are a constant $= \mu$. Also assume that the final length of the spring is l_0. (Actually, it will be somewhat longer that l_0, but if μ is small, this can be ignored.) Stuck? Look at Helping Questions 8 and 9.　　　　**Key 30**

PROBLEM IV

A block slides down a 1-m-high ramp, which is tilted at 30°. The coefficient of kinetic friction between the block and the ramp is $\mu_k = 0.4$.

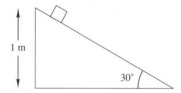

1. What is the block's speed at the bottom of the ramp? Stuck? Use Helping Question 10 and then Helping Question 11, if you need to.　　**Key 40**

PROBLEM V

A simple pendulum consisting of a mass M at the end of a string of length l is released from rest at an angle θ_0. A pin is located a distance $L < l$ directly below the pivot point.

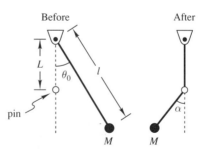

1. After the string hits the pin, what is the maximum angle α that the pendulum makes with respect to the vertical? Stuck? Use Helping Question 12 and then Helping Question 13, if you need to.　　**Key 19**
2. If instead the mass is given an initial velocity $\mathbf{v}_0$ as sketched here, what is the maximum angle α after the string hits the pin?　　**Key 12**

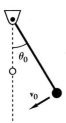

3. How is the answer to part (2) changed if v_0 is in the direction opposite to that shown in the figure? **Key 41**

PROBLEM VI

A small mass m starts from rest and slides from the top of a fixed sphere of radius r.

1. If the sphere is frictionless, at what angle θ from the vertical does the mass leave the surface? If you need a hint, use Helping Question 14. **Key 36**
2. Suppose there is friction between the mass and the sphere with $\mu_s = 0.1$. What is the minimum angle $\theta_{\min}$ at which the mass will start to slide along the sphere? **Key 46**
3. The mass is now placed just past this minimum angle and released. The coefficient of kinetic friction μ_k is small but nonzero. Does the mass fly off at a larger or a smaller θ than was found in part (1)? Assume that in both cases θ is defined with respect to the top of the sphere. **Key 14**

PROBLEM VII

The power of an automobile engine is usually measured in horsepower (hp) instead of watts (W). One horsepower equals 746 W. A typical automobile engine can sustain an output of 100 hp for a long period of time.

1. Find some clever way to estimate how much sustained power *you* can put out, say, for half an hour. Need an idea? Try Helping Question 15. **Key 9**

PROBLEM VIII

The gravitational force exerted on a planet by the sun is attractive, so the planet's potential energy is greater the farther from the sun it is. The planet moves in a slightly elliptical orbit, as indicated in the diagram.

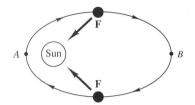

1. Is the planet moving faster at point A or point B? If you can't figure this out, turn to Helping Question 16. **Key 10**
2. During what part of the orbit is the sun doing positive work on the planet? Negative work? See Helping Question 17, if you're stuck here. **Key 3**
3. Check that when the sun is doing positive work the power $\mathbf{F} \cdot \mathbf{v}$ is positive; see Helping Question 18. **Key 15**

PROBLEM IX

The work done by a force depends on the reference frame from which the system is observed. The kinetic energy of a particle also depends on the reference frame. However, in all inertial reference frames the work–energy theorem $W = \Delta K$ holds. In this problem, you will verify these three statements for a sample system viewed from two different reference frames.

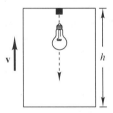

An elevator of height h moves upward at constant velocity v. Vibrations cause the elevator's light bulb to fall from its fixture at time $t = 0$. The bulb hits the floor at time t later. Call the mass of the bulb m.

1. One observer views the bulb from the elevator. In this frame, what is the initial kinetic energy K_0? The final kinetic energy K? The work W done by gravity? Since you are trying to verify the work–energy theorem, do not use any energy principles you have learned: calculate the work done from the definition $W = Fd$. **Key 11**
2. A second observer views the bulb from the ground. In this frame, what is the initial kinetic energy K_0? The final kinetic energy K? The work W done by gravity? Again, calculate the work done from the definition $W = Fd$. **Key 20**

Learning Guide 3   Work and Energy

3. Verify that the ground observer gets the same work–energy equation as the elevator observer. (The work–energy equation means $W = \Delta K$ expressed in terms of m, v, t, g, and h.) **Key 44**

HELPING QUESTIONS

1. What is the length of a unit vector? What is the angle between a unit vector and itself? Between two different unit vectors that point along different coordinate axes? Now you know enough to use the formula for the scalar product. **Key 39**

2. How can you use your results from part (2)? **Key 35**

3. The equation $y = h - gt^2/2$ is valid in the coordinate system that has the positive direction of the y-axis pointing upward. In this coordinate system what is the force **F**? What is the displacement **d** in time t? **Key 45**

4. Draw a force (free-body) diagram to determine the tension in the rope. **Key 43**

5. How is the rope tension related to force F? **Key 33**

6. This is a geometry question, not a physics question. The block finishes h ft closer to the ceiling; how much rope must be pulled past the top pulley on the left? **Key 28**

7. How can you use what you know about circular motion to get the kinetic energy of the mass? (Get rid of v^2!) **Key 37**

8. The only thing you know about the distance traveled is the frictional energy dissipated as the mass slows down and stops. How does this help? **Key 34**

9. What is the total energy of the system when the friction comes on? **Key 16**

10. How does the change in the sum of the kinetic energy plus the potential energy relate to the work done by friction? **Key 4**

11. What is the force of friction? Then what is the work done by friction? **Key 22**

12. What quantity is conserved throughout the motion? **Key 2**

13. What are the initial kinetic and potential energies, K_i and U_i? What are the final kinetic and potential energies K_f and U_f? Take the potential energy equal to zero at the lowest point of the motion. **Key 27**

14. Use conservation of energy to determine the speed v of the mass at angle θ. What is the normal force when the block just leaves the sphere? Your answer will give you a second equation relating v and θ. **Key 29**

15. How many stairs can you climb in half an hour? **Key 42**

16. Is the *total* energy of the system conserved? When is *kinetic* energy the greatest? **Key 24**

17. When the potential energy of the system is decreasing, is the sun doing positive or negative work on the planet? **Key 21**

18. What is the sign of $\mathbf{F} \cdot \mathbf{v}$ if the angle between **F** and **v** is less than 90°? Greater than 90°? **Key 13**

Notes: Conserved Quantities

Now that you've finished Learning Guides Nos. 2 and 3, you're able to solve mechanics problems in two quite different ways—by $\mathbf{F} = m\mathbf{a}$ and by energy conservation. As you progress in your study of mechanics, you'll develop an intuition for what types of problems can best be solved by $\mathbf{F} = m\mathbf{a}$ and what types of problems can be solved by energy conservation. But while you're learning, a good rule to work by is *always try to solve the problem first by energy conservation*. If the problem can be solved by both $\mathbf{F} = m\mathbf{a}$ and energy conservation, *it's almost always easier to solve it by energy conservation*. Let's look back at some examples.

Problem III, where the block slows down by friction, illustrates the point well. It's easy by energy methods once you get the idea. It's theoretically possible to solve this problem by $\mathbf{F} = m\mathbf{a}$, but first you'd need to know an equation describing the block's path. Thus, $\mathbf{F} = m\mathbf{a}$ is the hard way. Problem IV, about the block with friction, is a less clear-cut case. The helping questions suggested the combination of attack of using the work–energy theorem together with $\mathbf{F} = m\mathbf{a}$ to discover the work done by friction, but this is not really much easier than using only $\mathbf{F} = m\mathbf{a}$. Problem V, about the pendulum and the peg, is similar to Problem III: solvable by conservation of energy, practically impossible by $\mathbf{F} = m\mathbf{a}$. Problem VI, about the block sliding off the sphere, needed a combination attack. As its helping question pointed out, you needed two equations, one from $\mathbf{F} = m\mathbf{a}$ and one from energy conservation.

Soon you will learn about conservation of momentum, and after that you will learn about conservation of angular momentum. Since both momentum and angular momentum are vector quantities having three components each, you will have seven conserved quantities in all, quite a powerful arsenal. *You will be able to solve a wide range of physical problems by conservation laws alone. Some of the physical systems that you will study will be so complex that you won't possibly be able to identify all the forces, but you will still be able to get important results from the conservation laws alone.*

ANSWER KEY

1. $$\frac{\text{PE (spring)}}{\text{KE (mass)}} = \frac{R - l_0}{R}$$

2. Total energy

3. Bottom half; top half

4. (Change in mechanical energy) = (work done by friction). Both quantities are negative for this and all problems involving friction.

5. $d = 2nh$

6.

•	i	j	k
i	1	0	0
j	0	1	0
k	0	0	1

7. $W = \Delta K$, $\quad \dfrac{mg^2t^2}{2} = \dfrac{mg^2t^2}{2} - 0$

8. No, since $Fd = Wh$ for all choices of n. We know from the work–energy theorem that it must be this way.

Learning Guide 3   Work and Energy

9. Around $\frac{1}{6}$ horsepower, if your mass is 60 kg
10. Point A
11. $K_0 = 0$; $K = \frac{1}{2}m(gt)^2$; $W = mgh$
12. $$\alpha = \cos^{-1}\left[\left(\frac{1}{l-L}\right) \times \left(l\cos\theta_0 - L - \frac{v_0^2}{2g}\right)\right]$$
13. $+$; $-$
14. Larger
15. It checks: when the planet is going from B to A, the angle between $\mathbf{F}$ and $\mathbf{v}$ is *less* than 90°, so the power delivered is positive.
16. $$E_{\text{tot}} = \frac{kR}{2}(R - l_0) + \frac{1}{2}k(R - l_0)^2$$
17. $mg^2t^2/2$
18. $a_xb_x + a_yb_y + a_zb_z$
19. $$\alpha = \cos^{-1}\left(\frac{l\cos\theta_0 - L}{l - L}\right)$$
20. $K_0 = \frac{1}{2}mv^2$; $K = \frac{1}{2}m(-v + gt)^2$; $W = mg(h - vt)$
21. Positive
22. $$F_k = \mu_k N = \mu_k mg \cos\theta$$
$$W_k = \mathbf{F}_k \cdot \mathbf{d}_k = -\mu_k mgh \times (\cos\theta/\sin\theta)$$
23. $F = W/2n$
24. Yes; when the potential energy is the least
25. 24
26. No. l_0 cannot be zero for a real spring.
27. $K_i = 0$; $U_i = mgl(1 - \cos\theta_0)$
    $K_f = 0$; $U_f = mg(l - L)(1 - \cos\alpha)$
28. $2Nh$
29. From energy conservation:
$$\frac{1}{2}mv^2 = mgr(1 - \cos\theta).$$

From the normal force equaling zero at breakaway:
$$mg\cos\theta = \frac{mv^2}{r}.$$

30. $$\text{Distance} = \frac{k}{\mu mg}(R - l_0) \times \left(R - \frac{l_0}{2}\right)$$

31. $K_0 = 0$; $K = mg^2t^2/2$
32. 24 again
33. They are equal.
34. Energy lost to friction $= \mu mgd = $ total energy.
35. Multiply out $(a_x\mathbf{i} + a_y\mathbf{j} + a_z\mathbf{k}) \cdot (b_x\mathbf{i} + b_y\mathbf{j} + b_z\mathbf{k})$. You will get nine terms, and for each one of them you can use an entry from your multiplication table.
36. $\theta = \cos^{-1}(2/3) \simeq 48.2°$, independent of r and g.
37. $$\text{Force} = k(R - l_0) = \frac{mv^2}{R},$$
so
$$\frac{1}{2}mv^2 = \frac{1}{2}kR(R - l_0).$$
38. $$\frac{dW}{dt} = mg^2t,$$
which agrees with
$$\mathbf{F} \cdot \mathbf{v} = (-mg\mathbf{j}) \cdot (-gt\mathbf{j}) = mg^2t.$$
39. 1; 0°; 90°; so $\mathbf{i} \cdot \mathbf{i} = (1)(1)\cos 0° = 1$, etc.
40. $v = \sqrt{2gh(1 - \mu_k \cot\theta)}$
    $= 2.5$ m/s

41. The answer doesn't change.
42. Maybe 1 step per second, or 1800 steps. 5 steps is about 1 m. So you can lift your weight through 360 m in 1800 s. What's your power output?
43.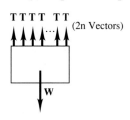
44. The ground observer gets an extra term, $-mgvt$, on each side of the equation.
45. $-mg$; $-gt^2/2$ (note both minus signs)
46. $\theta_{\min} = \tan^{-1}\mu = 5.7°$

learning guide 4

Linear Momentum and Collisions

Suggested Reading: Fishbane, Gasiorowicz, & Thornton, Chapter 8

PROBLEM I

Find the center of mass for the following collections of objects:

1. Two masses of 3 kg and 5 kg separated by a distance of 10 m **Key 10**
2. The following collection of masses:
   2 kg at $(0, 0, 0)$ m
   6 kg at $(0, 0, 4)$ m
   6 kg at $(0, 4, 0)$ m
   6 kg at $(4, 0, 0)$ m
   3 kg at $(1, 1, 1)$ m
   5 kg at $(-1, 0, 2)$ m
   2 kg at $(-3, 2, -7)$ m **Key 13**
3. A solid sphere of constant density **Key 30**
4. A thin spherical shell of constant density and thickness **Key 18**
5. A thin Hula-Hoop of constant density and thickness **Key 3**

PROBLEM II

A 198-lb man is carrying three coconuts each weighing one pound. He wants to cross a bridge that will hold 200 lb but not one ounce more. He decides to juggle the balls while crossing the bridge so that one ball will always be in the air.

1. Does he make it? If this doesn't seem like physics, look at Helping Questions 1 and 2. **Key 14**

PROBLEM III

A tennis ball of mass $m = 50$ g is hit at 30 m/s against a backboard. The ball returns with a speed of 20 m/s. In your analysis neglect the vertical motion of the ball.

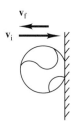

1. What is the change Δp in the ball's momentum? **Key 29**
2. What is the impulse J given to the ball? **Key 11**
3. Which of the graphs below could be a graph of the force on the backboard versus time? **Key 24**

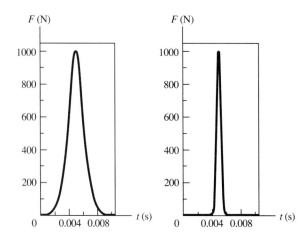

PROBLEM IV

A block with mass $m_1 = 10$ kg moving at 5 m/s collides with another block with mass $m_2 = 20$ kg moving the other way at 1 m/s. The two blocks stick together after the collision.

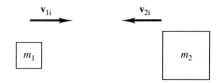

1. What is their common final velocity v_f? See Helping Question 3 if you're stuck. **Key 33**

The blocks collide again, this time elastically.

2. What are the final velocities $\mathbf{v}_{1f}$ and $\mathbf{v}_{2f}$? Assume that the outgoing blocks move away from the collision along the initial line of approach. If you need help somewhere along the line, use Helping Questions 4 and 5. **Key 34**

PROBLEM V

A Volkswagen weighing 1600 lb and a Mercedes weighing 4000 lb each moving at 44 ft/s (30 mi/h) enter an intersection from the north and south, respectively. They collide head on, and the resulting junk sticks together.

1. What is the final velocity of the junk? If you don't understand, use Helping Question 6. **Key 9**
2. What is the change in velocity experienced by a passenger in each car? **Key 35**

For the purpose of estimation, assume that the collision lasts 0.1 s and the deceleration of the cars is constant.

3. What is the magnitude of the deceleration of each car in terms of g, the gravitational constant? Take $g = 32$ ft/s^2. **Key 26**
4. If the Mercedes instead enters the intersection from the east, what are the magnitude and direction of the junk's velocity? **Key 22**

PROBLEM VI

In an elastic collision between two particles of *equal* mass, one of which is initially at rest, the recoiling particles always move off at right angles to one another. This may be proved directly from the equations of conservation of energy and momentum.

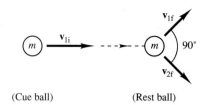

1. Write the *vector* equation for conservation of momentum. What does this imply about the relation between the velocities? **Key 21**
2. Write the equation for conservation of energy. What does this imply about the relationship between the speeds? **Key 2**
3. Take the scalar (dot) product of the relation between the velocities from part (1) with itself. **Key 5**
4. Compare the results of parts (2) and (3). What does this tell you about $\mathbf{v}_{1f} \cdot \mathbf{v}_{2f}$? What do you conclude? **Key 31**

PROBLEM VII

An "infinitely massive" Ping-Pong paddle moving with velocity $\mathbf{v}_{1i}$ hits a "massless" Ping-Pong ball at rest. The collision is elastic.

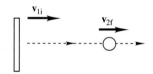

1. What is the final velocity of the ball, $\mathbf{v}_{2f}$? If you feel you have to resort to lengthy computations to get the answer, see Helping Questions 7 and 8 for a quicker way. **Key 6**

PROBLEM VIII

A block of mass m moving with velocity $\mathbf{v}_{1i}$ collides elastically with a stationary block, also of mass m. The massless spring of spring constant k is compressed for a short time during the collision.

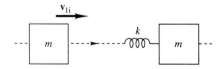

1. What is the maximum compression x of the spring? If you're lost, follow Helping Questions 9, 10, and 11. **Key 19**

Learning Guide 4   Linear Momentum and Collisions

PROBLEM IX

A billiard ball moving at a speed of 2.2 m/s strikes an identical stationary ball a glancing blow. After the collision, one ball is found to be moving at a speed of 1.1 m/s in a direction making a 60° angle with the original line of motion.

1. What are the magnitude and direction of the other ball's velocity?   **Key 8**
2. Is the collision inelastic?   **Key 1**

PROBLEM X

A fisherman in a canoe catches a very big fish. He observes that as he reels in 100 m of line, he (mass 70 kg) and the canoe (mass 25 kg) move 10 m toward the original position of the now quiescent fish, and away from a buoy at their original position. How heavy was the fish he caught? (No exaggeration. Please!). Assume that the fish and the boat are at the same height (i.e., both at the surface) and ignore the frictional interaction with the water and the size of the canoe. Stuck? Check out Helping Question 12.   **Key 20**

HELPING QUESTIONS

1. Describe in words the motion of the center of mass of the man-coconuts system as it moves across the bridge.   **Key 7**
2. What is Newton's second law for a *system* of particles? Can you use this to find the normal force acting upward of the man's feet? How does this relate to the force of the man's feet acting downward on the bridge?   **Key 17**
3. What quantity is conserved? What equation relating the masses and velocities does this conservation law give?   **Key 28**
4. What quantities are conserved? What equations relating the masses and the velocities do the conservation laws give?   **Key 25**
5. Solving the two equations given by the conservation laws for v_{1f} and v_{2f} in terms of m_1, v_{1i}, m_2, and v_{2i} is a little tricky. It's easier if you go contrary to general practice and plug in numbers for the givens *before* solving the equations. If you're interested in a general solution, see Fishbane, Gasiorowicz, & Thornton, Section 8-4.
6. What quantity is conserved? What equation relating the masses of the two cars and the initial and final velocities does this equation give?   **Key 4**
7. How would an observer moving from left to right at speed v_{1i} with the paddle describe the collision? At what speeds would he see the ball, before and after the collision?   **Key 15**
8. Since the observer moving with the paddle sees the ball go off to the right at speed v_{1i}, what ball speed does an observer in the original frame see?   **Key 23**
9. Think about the instant of maximum compression. What is

the *relative velocity* of the two blocks at that instant? What is the velocity of each block at the instant of maximum compression? **Key 16**

10. What is the total energy of the system at the instant of maximum compression in terms of the variables given in the problem? **Key 12**

11. What is the initial kinetic energy of this system? **Key 32**

12. Where is the center of mass of the man, rowboat, and fish after the fish has been reeled in? **Key 27**

Notes

Number of Unknowns and Number of Equations

As you do the problems in these Learning Guides, you may sometimes get the feeling that the whole business is getting out of hand—there are so many variables that you can't possibly keep track of all of them.

An excellent way to organize your thoughts is to divide the variables into two groups, knowns and unknowns. A known variable doesn't necessarily have to be known in the sense that you know its numerical value. A known variable is a variable you're allowed to leave in the answer. An unknown variable is a variable you want to express in terms of the known variables. The reason that this distinction is so useful is that *you need to find as many different equations as there are unknowns.* This is only a rule of thumb, not a theorem, but it works very well. Let's consider as examples three of the problems in this Learning Guide:

	Problem	Knowns	Unknowns	Equations
IV	Inelastic collisions	m_1, v_{1i}, m_2, v_{2i}	v	Conservation of momentum in the x-direction
VIII	1-D elastic collisions	m_1, v_{1i}, m_2, v_{2i}	v_{1f}, v_{2f}	Conservation of momentum in the x-direction; conservation of energy
VI	2-D elastic collisions	m_1, v_{1i}, m_2, v_{2i}	v_{1f}, v_{2f}, θ	Conservation of momentum in the x- and y-directions (two equations); conservation of energy

Suppose you wanted to solve the two-dimensional elastic collision completely—that is, you wanted to find not only the angle θ between the paths of the recoiling particles, but also the angles θ_1 and θ_2 that the paths make with the x-axis. Then you would have four unknowns, v_{1f}, v_{2f}, θ_1, and θ_2, and no matter how you floundered about using the three conserved quantities p_x, p_y, and E, you could not solve the problem. This has a physical interpretation: the recoiling

Learning Guide 4   Linear Momentum and Collisions

velocity vectors are not determined by p_x, p_y, and E alone; they depend on the details of the forces in the collisions.

You might want to examine a few of your solutions to problems in previous Learning Guides to verify (number of unknowns) = (number of equations). *Notice for a given problem the number of unknowns is not fixed.* You might introduce an unknown frictional force F_k while solving a force problem, while someone else might immediately write down $\mu_k N$. You would have an extra unknown but you would also have an extra equation $F_k = \mu_k N$. Another good point about this way of organizing your thoughts is that you can see right away where the physics ends and the algebra begins: *the physics ends when you have found enough equations to determine your unknowns.*

Choice of Reference Frame

Turning to another subject, Newton's laws (and thus the conservation laws you have learned) are valid in any *inertial* reference frame. But for a given situation, as you saw in Problem VII, certain reference frames may be much more useful than others. What reference frame was used in Problem VII? *The center-of-mass frame.* What's so special about the center-of-mass frame? (See Chapter 8 in Fishbane, Gasiorowicz, & Thornton for the definition of center of mass and the center-of-mass velocity **V**.)

First, let's look at what happens to the kinetic energy,

$$K = \tfrac{1}{2}m_1 v_1^2 + \tfrac{1}{2}m_2 v_2^2,$$

in a one-dimensional collision. The kinetic energy can be divided into two types: kinetic energy *of* the center of mass, here called K_{CM}, and kinetic energy *with respect to* the center of mass, here called $K_{internal}$. Let **v** represent the velocity of the center of mass. Then the kinetic energy of the center of mass is

$$K_{CM} = \tfrac{1}{2}(m_1 + m_2)V^2,$$

and the kinetic energy with respect to the center of mass is

$$K_{internal} = \tfrac{1}{2}m_1(\mathbf{v}_1 - \mathbf{V}_{CM})^2 + \tfrac{1}{2}m_2(\mathbf{v}_2 - \mathbf{V})^2.$$

*Nothing happens to K_{CM} in a collision, since **V** remains constant. $K_{internal}$ always decreases to zero as the particles collide and then returns to some fraction of its initial value* depending on how elastic the collision is. If you didn't use the center-of-mass frame to solve Problem VIII, go back and write the energy equations as viewed from a frame moving to the right with speed $v_{1i} (= V)$. Notice that in this frame it is easier to get the spring compression x because all and not just some of the kinetic energy is momentarily transferred to potential energy of the spring. *All types of collisions look simple when you move along with the center of mass,* in such a way that the center of mass is fixed in your frame:

Inelastic collision

(Equal but opposite momenta)

(Just stick together)

Partly elastic collisions

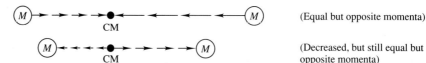

(Equal but opposite momenta)

(Decreased, but still equal but opposite momenta)

Elastic collisions

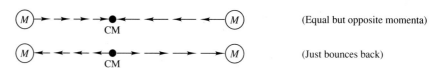

(Equal but opposite momenta)

(Just bounces back)

The Helping Questions mentioned two ways to solve Problem IV. See if you can do it yet a third way by using the center-of-mass frame!

If you keep these two ideas in mind—number of equations versus number of unknowns and choice of reference frame—the problems in the next Learning Guide will be quite a bit easier.

ANSWER KEY

1. No
2. $\frac{1}{2}mv_{1i}^2 = \frac{1}{2}mv_{1f}^2 + \frac{1}{2}mv_{2f}^2$,

   so $v_{1i}^2 = v_{1f}^2 + v_{2f}^2$.

3. At the center
4. Linear momentum is conserved, so

   $m_1 v_{1i} + m_2 v_{2i} = (m_1 + m_2)v_f$.

5. $v_{1i}^2 = v_{1f}^2 + 2\mathbf{v}_{1f} \cdot \mathbf{v}_{2f} + v_{2f}^2$
6. $v_{2f} = 2v_{1i}$
7. The center of mass moves almost horizontally across the bridge at whatever speed the man is walking. It bobs up and down a little, because of the juggling of the coconuts and the "bounce" in the man's step. However, there cer- tainly is some time when the vertical acceleration of the center of mass is zero.
8. 1.9 m/s, 30° from $\mathbf{v}_i$
9. $v = \dfrac{W_1 v_{1i} + W_2 v_{2i}}{W_1 + W_2}$

   $= 12.9$ mi/h $= 18.9$ ft/s

   in the direction of the Mercedes.
10. $3\frac{3}{4}$ m from the 5 kg mass on the line segment between the two masses
11. -2.5 kg·m/s ($\mathbf{j}$ points to the left)
12. $\frac{1}{4}mv_{1i}^2 + \frac{1}{2}kx^2$
13. $\left(\frac{8}{15}, \frac{31}{30}, \frac{23}{30}\right)$ m
14. No. It might have collapsed even without the coconuts. See Helping Question 1.
15. He would say that a ball mov-

/ Learning Guide 4   Linear Momentum and Collisions

ing at speed v_{1i} to the left hits a solid, stationary wall. So, since the collision is elastic, the ball just bounces back with speed v_{1i}.

16. 0; $v_{1i}/2$
17. $\mathbf{F}_{ext} = M\mathbf{a}$. Yes, when there is no vertical acceleration the normal force is exactly 201 lb. They are equal.
18. At the center
19. $x = \sqrt{\dfrac{m}{2k}} v_{1i}$
20. 10.56 kg
21. $m\mathbf{v}_{1i} = m\mathbf{v}_{1f} + m\mathbf{v}_{2f}$, so $\mathbf{v}_{1i} = \mathbf{v}_{1f} + \mathbf{v}_{2f}$.
22. 33.7 ft/s; 68° west of south
23. $2v_{1i}$
24. The first
25. Conservation of momentum gives

    $m_1 v_{1i} + m_2 v_{2i} = m_1 v_{1f} + m_2 v_{2f}.$

    Conservation of energy gives

    $\tfrac{1}{2} m_1 v_{1i}^2 + \tfrac{1}{2} m_2 v_{2i}^2$
    $= \tfrac{1}{2} m_1 v_{1f}^2 + \tfrac{1}{2} m_2 v_{2f}^2.$

26. $a = 7.8g$ for the Mercedes and $a = 19.7g$ for the Volkswagen.
27. At the center of the boat, 10 m from the buoy
28. Linear momentum is conserved, so

    $m_1 v_{1i} + m_2 v_{2i} = (m_1 + m_2) v_f.$

29. $m(v_f - v_i) = -2.5$ kg·m/s (taking the positive direction to the right).
30. At the center
31. $\mathbf{v}_{1f} \cdot \mathbf{v}_{2f} = 0$, so either one of the balls is stopped, or they're moving at right angles.
32. $m v_{1i}^2 / 2$
33. $v_f = \dfrac{m_1 v_{1i} + m_2 v_{2i}}{m_1 + m_2} = 1$ m/s to the right.
34. $\mathbf{v}_{1f} = -3$ m/s; $\mathbf{v}_{2f} = 3$ m/s
35. 25 ft/s for the Mercedes passenger and 63 ft/s for the Volkswagen passenger

learning guide 5

...ional Motion

Suggested Reading: Fishbane, Gasiorowicz, & Thornton, Chapters 9 and 10

VECTOR WARM-UPS—THE VECTOR PRODUCT

The vectors **a**, **b**, and **c** all lie in the plane of the paper. Give the magnitude and direction of the following vector products:

1. **a** × **b**. See "A Closer Look" in Section 10-1 of Fishbane, Gasiorowicz, & Thornton to learn about vector products.   **Key 10**
2. **a** × **c**   **Key 3**
3. **a** × **a**   **Key 41**
4. **b** × **a**   **Key 36**

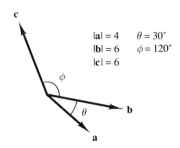

|**a**| = 4 $\theta = 30°$
|**b**| = 6 $\phi = 120°$
|**c**| = 6

There is another method that is occasionally more convenient for calculating vector products. You will discover the method in parts (5) and (6) and then use it in part (7).

5. **i**, **j**, and **k** are the unit vectors drawn in the diagram. Fill in the multiplication table below:

×	i	j	k
i			
j			
k			

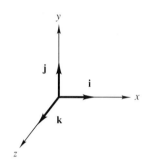

   If you need help, turn to Helping Question 1. **Key 26**

6. In a certain fixed coordinate system,

$$\mathbf{a} = a_x\mathbf{i} + a_y\mathbf{j} + a_z\mathbf{k} \quad \text{and} \quad \mathbf{b} = b_x\mathbf{i} + b_y\mathbf{j} + b_z\mathbf{k}.$$

   What is $\mathbf{a} \times \mathbf{b}$ in terms of the unit vectors and the components of **a** and **b**? If both $a_z = 0$ and $b_z = 0$, what does your answer reduce to? See Helping Question 2. **Key 35**

7. The vectors from parts (1) through (4) are given in the sketch in terms of their components in a rectangular coordinate system. Check your answers to parts (1) to (4) using the new method.

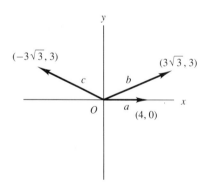

PROBLEM I

The rigid body shown in the figure consists of four 10-kg spheres connected by four light rods. Treat the spheres as point particles and neglect the mass of the rods.

Learning Guide 5 Rotational Motion

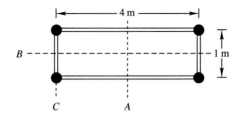

1. Which is greater, the rotational inertia about axis A or the rotational inertia about axis B? What are their exact values? **Key 24**
2. Use the parallel-axis theorem to calculate the rotational inertia about axis C. Check your answer by calculating the rotational inertia about axis C from the definition. **Key 6**

PROBLEM II

A string wrapped around a solid cylinder of mass M and radius R is pulled vertically upward to prevent the cylinder from falling as it unwinds the string. (That is, the center of mass of the cylinder does not move.)

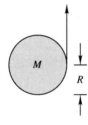

1. What is the tension in the string? If you don't agree with the key, use Helping Question 3. **Key 30**
2. If the cylinder is initially at rest, how much string is unwound after a time t? If, after a good effort, you're stuck, use Helping Questions 4 and 5. **Key 39**

PROBLEM III

In the apparatus shown in the sketch, both blocks accelerate as a result of the force of gravity on m_2. The coefficient of kinetic friction between m_1 and the table is μ_k. The pulley has frictionless bearings, mass M, and radius R.

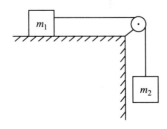

1. The blocks are initially at rest. At time t, through what distance y has m_2 moved? Use Helping Questions 6, 7, and 8 if you need to. **Key 21**

PROBLEM IV

After the block of mass m_2 in Problem III has fallen a distance y from rest, it has speed v.

1. Use the work–energy theorem to express v in terms of $m_1, m_2, \mu_k, g, y, I,$ and R. Use as many of Helping Questions 9 to 11 as are necessary. **Key 31**

PROBLEM V

A diver does a two-and-a-half front somersault from a 3-meter board. While she is in her tuck, the rotational inertia of her body is $(4 \text{ slugs})(1 \text{ ft})^2$. She is spinning at a rate of one complete revolution per second. After she extends her body to enter the water, her rotational inertia is $(4 \text{ slugs})(2 \text{ ft})^2$.

1. How long would it take her to complete one revolution at her new rate of rotation? If you're confused, try Helping Question 12. **Key 32**
2. Did her rotational kinetic energy increase or decrease? How do you account for the change? If you need a hint for the first question, use Helping Question 13. **Key 5**

PROBLEM VI

A disk of wood is mounted on frictionless bearings, leaving it free to rotate about its center. The mass of the disk is 1 kg and its radius is 10 cm. A 5-g bullet traveling at 500 m/s lodges in the disk 5 cm above its center, as indicated on the diagram. The disk starts to rotate. Assume that the displacement of the wood is negligible, and ignore the rotational inertia due to the bullet.

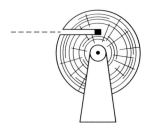

1. How many seconds does it take for the disk to make one revolution after the bullet strikes? See Helping Questions 14 and 15. **Key 22**

Learning Guide 5 Rotational Motion

PROBLEM VII

The "sweet spot" of a tennis racket is the spot where the player feels the least vibration through his wrist and forearm when he hits the ball. In this problem you will ignore many factors but still make an accurate computation of the location of the sweet spot. Assume that there are no external forces on the racket shown in the sketch. The sweet spot of the racket is indicated by a cross. When the ball hits the stationary racket on the sweet spot, the velocity of the center of the grip right after the collision is zero (so that if a player were holding it, he wouldn't feel the grip jar forward or backward during the collision).

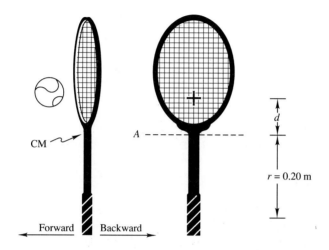

1. If the ball hits the racket *beneath* the sweet spot, will the initial velocity of the grip be forward or backward? Use Helping Question 16 if you don't know. **Key 37**
2. If the ball hits the racket *above* the sweet spot, will the initial velocity of the grip be forward or backward? **Key 2**
3. What is the distance d between the center of mass and the sweet spot? Stuck? Use Helping Questions 17 and 18. **Key 9**

PROBLEM VIII

A pool cue strikes a pool ball which is sitting on a level pool table with a coefficient of kinetic friction μ_k. The ball is given an initial speed of v_0 with no spin.

1. How fast is the ball moving when it begins to roll without slipping? Even though the center-of-mass frame is a *noninertial reference frame*, $\tau = I\alpha$ still holds in this frame—see Section 9-7 of Fishbane, Gasiorowicz, & Thornton. If you need more hints, use Helping Questions 19, 20, and 21. **Key 42**

PROBLEM IX

In Problems I through VIII, you did not get to use the *vector* equation $\tau = d\mathbf{L}/dt$. In part (2) of this problem you will!

A heavy wheel rotates with no friction and a very high angular velocity ω about a massless shaft. One end of the shaft is fixed to the wall, but the shaft is free to pivot about this point. The other end of the shaft is held by a physics student. The wheel has mass M and rotational inertia I about the shaft.

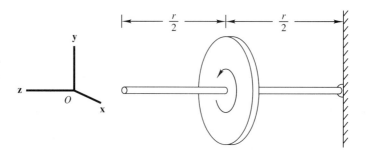

1. The student wants to hold the shaft fixed. What force must she apply to the end of the shaft? **Key 15**
2. The student now wishes to raise the shaft by imparting a velocity $\mathbf{v} = v\mathbf{j}$ to the end of the shaft. What force should she apply to the end of the shaft? Be sure to specify the direction of the force! You may assume that since ω is so large, the angular momentum $\mathbf{L}$ about the pivot always points along the direction of the shaft. For a hint, look at Helping Question 22. **Key 11**

HELPING QUESTIONS

1. What is the length of a unit vector? What is the angle between a unit vector and itself? Two different vectors? How can you figure out the *direction* of a cross product? **Key 17**
2. How can you use your multiplication table? **Key 33**
3. Is the cylinder's center of mass moving? Then what must be the sum of the external forces on the cylinder? **Key 7**
4. What is the angular acceleration of the cylinder? **Key 23**
5. What is the rotational analogue of the translational kinematic formula $\Delta x = v_0 t + \frac{1}{2}at^2$? **Key 25**
6. Draw force (free-body) diagrams for each block and for the pulley. Call the tension in the horizontal part of the string T_1 and the tension in the vertical part of the string T_2. What is the torque τ on the pulley in terms of T_1 and T_2? What are T_1 and T_2 in terms of μ_k, m, g, and a, the acceleration of the blocks? **Key 1**
7. Can you relate the torque on the pulley to its angular acceleration? Can you relate the angular acceleration to the linear acceleration of the blocks? Now

can you solve for the acceleration a? **Key 12**
8. What did you get for the acceleration a? What kinematic formula would be appropriate to get the distance y? **Key 16**
9. What is the change in potential energy? **Key 13**
10. What is the change in kinetic energy? **Key 4**
11. How much energy is lost to friction? **Key 19**
12. What physical quantity is conserved throughout the dive? Since I increases by a factor of 4, what happens to ω? **Key 34**
13. What is the formula for rotational kinetic energy in terms of I and ω? **Key 18**
14. What physical quantity is conserved throughout the collision? Why? **Key 8**
15. What was the angular momentum about the center of the wheel before the collision? **Key 14**
16. Consider an extreme case: what would happen if the ball hit the racket at the level of the center of mass? **Key 27**
17. The racket recoils with velocity v_{CM} and angular velocity ω about its center of mass. What relation must hold between v and ω for the center of the grip to have zero velocity immediately after impact? **Key 40**
18. If the linear momentum the ball transfers to the racket is p, what is the racket's angular momentum about its center of mass? **Key 29**
19. What is the condition required for rolling without slipping? **Key 38**
20. What are the forces on the ball? Then what is $v(t)$? **Key 43**
21. What are the torques on the ball? Then what is $\omega(t)$? **Key 28**
22. What is $d\mathbf{L}/dt$ in terms of I, ω, r, and $\mathbf{v}$? **Key 20**

Notes—Analogies

You're probably finding rotational motion to be the hardest subject to understand in this course so far. It's no wonder. First, you've been introduced to a whole slew of physical concepts all at once—six, to be exact, θ, ω, α, τ, $\mathbf{L}$, and I. Worse yet, you have to be very precise when you write down equations describing rotational motion. Torques have to be defined with respect to a point (not an axis!) and rotational inertias have to be defined with respect to an axis (not a point!). You have to define your variables very carefully even when you write down an innocent looking equation like $v = \omega r$. You had to use this equation to solve Problem IV. There it was true by definition: the linear velocity of a particle in a rigid body rotating with angular velocity ω about an axis a distance r from the particle is ωr. You also had to use this equation in Problems VII and VIII. But there the variables were defined differently and $v = \omega r$ held only under certain conditions.

Is there anything to guide you through this maze of subtleties? Yes—*analogies* with what you've learned previously! You should be able to reproduce Table 9-2 in Fishbane, Gasiorowicz, & Thornton without any strain. $\tau = I\alpha$ shouldn't seem like a completely new equation to you—it's just the rotational analogue of $F = ma$. What are the linear analogues of the following equations?

$$\theta = \omega_0 t + \tfrac{1}{2}\alpha t^2$$
$$\mathbf{L} = I\omega$$
$$P = \tau \cdot \omega$$
$$E = \tfrac{1}{2}I\omega^2$$

You might get the impression from these tables that the equations you need to solve physics problems fall roughly into two categories—equations among linear variables and equations among rotational variables. *There is really a third category—equations relating linear variables to rotational variables. Every analogy between a rotational variable and a linear variable* (except for the analogy between θ and x) **corresponds to an equation relating the two variables and r.** For kinematic variables, the r is on the rotational variable's side:

$$v = \omega r$$
$$a = \alpha r,$$

where a means **tangential** acceleration. For dynamical variables the r is on the linear variable's side:

$$\boldsymbol{\tau} = \mathbf{r} \times \mathbf{F}$$
$$\mathbf{L} = \mathbf{r} \times \mathbf{p}$$

All equations work out dimensionally because there are two r's involved in the analogy between mass and rotational inertia:

$$I = (\text{constant})mr^2.$$

Of course, you can't just write down a bunch of equations "by analogy" and expect to solve a physics problem: you have to examine each problem separately and carefully define your variables so that your equations make sense. But the analogies serve as an extremely useful guide. As an example, let's look at the solution suggested by the Helping Questions to Problem III:

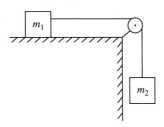

$T_1 - \mu_k m_1 g = m_1 a$	(linear)
$m_2 g - T_2 = m_2 a$	(linear)
$\tau = R(T_2 - T_1)$	(both)
$\tau = I\alpha$	(rotational)
$a = R\alpha$	(both)
$y = at^2/2$	(linear)

(six unknowns: $T_1, T_2, a, \alpha, t,$ and y).

ANSWER KEY

1.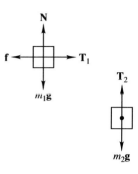

   $\tau = R(T_2 - T_1)$

   (into the page);

   $m_2 g - T_2 = m_2 a$;

   $T_1 - m_1 \mu_k g = m_1 a$.

2. Forward
3. 12, out of the page
4. $K_f - K_i = \frac{1}{2}m_1 v^2 + \frac{1}{2}I\omega^2 + \frac{1}{2}m_2 v^2$
5. Kinetic energy decreases; the woman does *negative internal work* as she lets her arms and legs "fly out."
6. $I_C = 320$ kg·m^2
7. No; 0
8. The *total angular momentum* of the bullet plus the disk about the center of the disk is conserved because there are no external torques about the center of the disk.
9. $d = \dfrac{I}{mr} \cong 11$ cm
10. 12, out of the page
11. $\dfrac{I\omega v}{r^2}\mathbf{i} + \dfrac{mg}{2}\mathbf{j}$ plus an arbitrary force in the z-direction
12. $\tau = I\alpha$; yes: $a = r\alpha$
13. $U_f - U_i = -m_2 g y$
14. $L = pr_\perp = 0.125$ kg·m^2/s, where $r_\perp$ is the distance of closest approach between the bullet's trajectory and the center of the wheel.
15. $mg/2$ (up) $= mg/2\mathbf{j}$ (plus an arbitrary force in the z-direction)
16. $a = \dfrac{m_2 g - m_1 \mu_k g}{m_1 + m_2 + I/R^2}$;

    $y = v_{0y} t + \frac{1}{2} a t^2$
17. 1; 0; 90°; by the right-hand rule
18. $I\omega^2/2$
19. $\mu_k m_1 g y$
20. $I\omega v / r$
21. $y = \dfrac{1}{2}\left(\dfrac{m_2 g - m_1 \mu_k g}{m_1 + m_2 + M/2}\right) t^2$
22. About 0.25 s
23. $2g/R$
24. $I_A > I_B$; $I_A = 160$ kg·m^2 and $I_B = 10$ kg·m^2
25. $\Delta \theta = \omega_0 t + \alpha t^2 / 2$
26.

×	i	j	k
i	0	k	−j
j	−k	0	i
k	j	−i	0

27. The racket would move backward without spinning.
28. $\mu_k mgr$; $\omega(t) = \mu_k mgrt/I$
29. pd
30. Mg
31. $v = \sqrt{\dfrac{(m_2 - m_1 \mu_k)}{(m_1 + I/R^2 + m_2)} 2gy}$
32. 4 s
33. Multiply out

    $\mathbf{a} \times \mathbf{b} = (a_x \mathbf{i} + a_y \mathbf{j} + a_z \mathbf{k})$
    $\times (b_x \mathbf{i} + b_y \mathbf{j} + b_z \mathbf{k})$.

    You'll get nine terms. For each one, you can use an entry from your multiplication table.
34. The woman's angular momentum about her center of mass; ω decreases by a factor of 4.

35. $\mathbf{a} \times \mathbf{b} =$
$(a_y b_z - a_z b_y)\mathbf{i}$
$+(a_z b_x - a_x b_z)\mathbf{j}$
$+(a_x b_y - a_y b_x)\mathbf{k},$
which reduces to
$\mathbf{a} \times \mathbf{b} = (a_x b_y - a_y b_x)\mathbf{k}.$
36. 12, into the page
37. Backward
38. $v = \omega r$
39. gt^2
40. $v_{CM} = \omega r$
41. 0
42. $v = \dfrac{v_0}{(1 + I/mr^2)} = \dfrac{5}{7} v_0$
43.

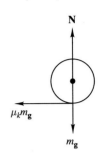

$v(t) = v_0 - \mu_k g t$

learning guide 6

Statics

Suggested Reading: Fishbane, Gasiorowicz, & Thornton, Chapter 11

PROBLEM I

A simple spring scale is attached to two 10-lb weights as shown. What does the scale read? See Helping Questions 1 and 2. **Key 31**

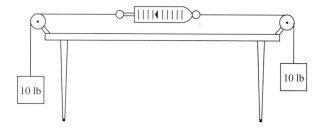

PROBLEM II

A patient's leg is supported by a traction apparatus consisting of a pulley and a suspended weight having mass $m = 2.5$ kg as shown in the accompanying sketch. Find the magnitude and direction of the force exerted on the patient's foot. See Helping Question 3. **Key 3**

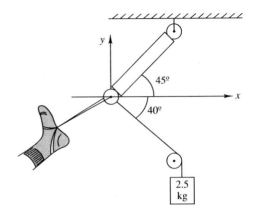

PROBLEM III

A massless board is nailed down to two vertical pillars spaced 3 m apart. A block of mass $m = 30$ kg rests on the board.

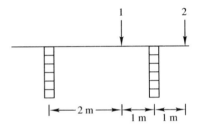

1. If the block is in position 1, what are the forces exerted on the board by the pillars? If you need them, use Helping Questions 4 and 5. **Key 19**
2. If the block is in position 2, what are the forces exerted on the board by the pillars? **Key 24**

PROBLEM IV

In terms of h, r, and W, what is the minimum force F necessary to start the wheel moving over the bump? Stuck? Try Helping Questions 6, 7, and 8.
Key 25

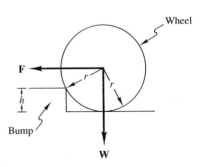

Learning Guide 6   Statics

PROBLEM V

The device shown in the sketch is a balance for comparing the weights of objects. The two objects to be compared are placed on the two trays hanging by wires from the ends of the arms. If the two objects have exactly the same weight, then the meter points straight up. If m_1 is different from m_2, then the meter is deflected by an angle ϕ. (The arms each have length r.)

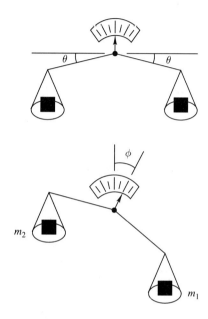

1. If the arms form a straight line so that the angle θ is zero, does the balance work? If you don't agree with the answer in the key, look at Helping Question 9. **Key 13**
2. If $\theta > 0$, at what angle ϕ does the balance reach equilibrium in terms of m_1, m_2, and θ? If you need them, use Helping Questions 10 and 11. **Key 7**

PROBLEM VI

The rod shown in the figure is tied to the wall by strings. It is perfectly horizontal and in static equilibrium.

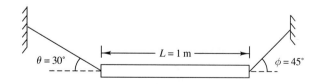

1. Is the rod of uniform density and diameter? If you need to write equations to answer this, use Helping Question 12. **Key 5**

2. Where is the center of gravity of the rod? Stuck? Try Helping Questions 13 and 14. **Key 22**

PROBLEM VII

A uniform-density plank of length l and mass m rests on the ground and on a frictionless roller (not shown) at the top of a wall of height h.

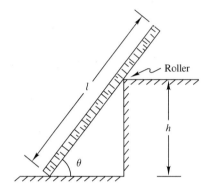

1. Find the direction and magnitude of the force exerted by the roller on the plank. Stuck? Try Helping Questions 15 and 16. **Key 15**
2. The plank slips for $\theta < \theta_0$. What is the coefficient of friction μ_s between the plank and the ground? Use Helping Question 17 if you really need it. **Key 14**

PROBLEM VIII

To calculate all the forces on an *underdetermined system* that is in static equilibrium, you have to use the laws of *elasticity* as well as the static equilibrium conditions $\mathbf{F}_1 + \mathbf{F}_2 + \cdots = 0$ and $\boldsymbol{\tau}_1 + \boldsymbol{\tau}_2 + \cdots = 0$. A simple example is shown in the sketch. Ignore gravity.

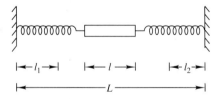

1. By counting equations and unknowns, show that the tensions T_1 and T_2 cannot be determined from $\mathbf{F}_1 + \mathbf{F}_2 + \cdots = 0$ and $\boldsymbol{\tau}_1 + \boldsymbol{\tau}_2 + \cdots = 0$. **Key 27**
2. Now suppose you know the unstretched spring lengths l_1, l_2 of the springs, the length l of the rigid rod, and the distance L between the walls. You also know the elastic properties of the springs. If spring 1 is stretched a distance x_1, then $T_1 = k_1 x_1$, where k_1 is the spring constant of spring 1;

Learning Guide 6   Statics

similarly, $T_2 = k_2 x_2$. Now find T_1 and T_2 (in terms of l_1, l, l_2, L, k_1, and k_2). Use Helping Question 18 if you need another equation.   **Key 2**

A car with four wheels is underdetermined by one equation. To gain some insight, consider the following simple model for the car. Suppose that it consists of a rigid block of mass m (of perhaps nonuniform density) representing the car body and four springs representing the tires. The distance from the front to the back tires is x and the distance from the left to the right tires is z. The center of gravity of the car is over a point x_{CG} behind and z_{CG} to the right of the front left wheel. The springs all have spring constant k and the same uncompressed length.

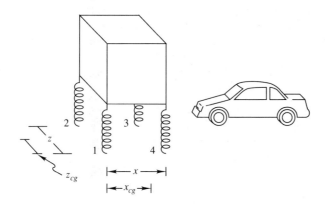

3. Write down a system of equations that are sufficient to determine the forces on the car body from each of the four tires. Don't bother solving the equations. Give this a good shot before looking at Helping Questions 19 and 20.   **Key 20**

In this problem, the bodies were assumed to be perfectly rigid and the springs were assumed to only expand and compress. In more advanced treatments of static equilibrium, the twisting and bending of both the body and the springs can also be taken into account.

HELPING QUESTIONS

1. What is the tension in the strings that hold up the weights?   **Key 11**

2. In "normal" operation the scale is usually held from above by a string attached to its top with the weight to be suspended using a second string attached to the bottom of the scale. In that case what would the tensions be for a 10-lb weight?   **Key 1**

3. What are the $\sum F_x$ and $\sum F_y$ acting on the central pulley?   **Key 10**

4. Call the forces from pillars 1 and 2, respectively, F_1 and F_2, where positive values for F_1 and F_2 mean that the force is up. What does translational equilibrium imply for this situation?   **Key 30**

5. Pick a convenient pivot point. What does $\tau_1 + \tau_2 + \cdots = 0$ say for this situation?   **Key 4**

6. How is this a static equilibrium question? **Key 9**
7. What is the most convenient point to chose as the pivot point? Why? **Key 6**
8. What is $\tau_1 + \tau_2 + \cdots = 0$ in terms of the variables given? **Key 18**
9. When the balance is deflected at an angle ϕ, what are the moment arms of the two forces about the pivot point? Does their *ratio* depend on ϕ? **Key 21**
10. Define ϕ to be positive when the balance is deflected to the right. When the balance is deflected through an angle ϕ, what is the moment arm about the pivot of $m_1\mathbf{g}$? Of $m_2\mathbf{g}$? **Key 28**
11. Do you remember any trigonometric identities for $\cos(\theta + \phi)$ and $\cos(\theta - \phi)$? After using these identities, solve for ϕ by first separating the terms containing $\sin\phi$ from the terms containing $\cos\phi$. **Key 17**
12. If the center of gravity were at the midpoint of the rod, what relation would have to hold between θ and ϕ for the horizontal bar to be in equilibrium? **Key 32**
13. Call the tensions in the strings T_1 and T_2. What does Newton's law $\mathbf{F}_1 + \mathbf{F}_2 + \cdots = 0$ say in terms of the variables given? **Key 8**
14. Take the pivot point at the leftmost point in the rod. Call the distance from the pivot point to the center of gravity x. What does $\tau_1 + \tau_2 + \cdots = 0$ say in terms of the variables given? **Key 33**
15. The force must be perpendicular to the plank. Why? **Key 29**
16. Draw a good force (free-body) diagram of the plank. What is the trick to get the force exerted by the roller with only one equilibrium equation? **Key 12**
17. What is the relationship between μ_s and the horizontal and vertical forces (H and V) at the bottom of the plank? **Key 16**
18. You should have found $T_1 = T_2$, $T_1 = k_1 x_1$, and $T_2 = k_2 x_2$. There is a fourth equation that comes from the *geometry* of the system. What is it? **Key 26**
19. Find eight equations to determine eight unknowns T_1, T_2, T_3, T_4, y_1, y_2, y_3, and y_4, where the y's are the compressions of the springs. So that your answer matches the key, take torques about the center of gravity. **Key 20**
20. Find the eighth equation by looking at the *geometry* of the system. How many points determine a plane? **Key 23**

Notes: Review

You'll probably agree that this is a good time to review. These notes will outline a four-step strategy for attacking physics problems, using Problems II to VII as examples. Most of the points in this strategy have been touched upon in earlier notes. You should think about whether or not this outline describes your basic strategy for attacking problems.

Step 1: Think about the Problem a Little Bit before Writing Equations.
Equations have a way of confusing the issue, especially if you don't have a physical understanding of the situation. If you decide that the problem is a forces prob-

Learning Guide 6   Statics   59

lem, draw a force (free-body) diagram to make sure you're not missing anything. You've had enough practice solving physics problems that you can guess certain features of the answer immediately. In Problem III, for example, you should be able to see right away that if the mass is right above pillar 1, all the weight is on pillar 1, and if the mass is right in between the two pillars, the weight is shared equally. You might even be able to suspect the most interesting part of the problem—that if the mass is not between the two pillars, then the force on one of the pillars is greater than its weight while the force on the other is negative. If you decide that the problem is a static equilibrium problem, then you should understand the problem well enough physically to be able to choose a convenient pivot point before moving on to step 2.

Step 2: Find As Many Equations As You Have Unknowns. Most of the time these equations are not hard to find—they come from conservation of energy, conservation of momentum, the components of $\mathbf{F} = m\mathbf{a}$, the components of $\tau_1 + \tau_2 + \cdots = 0$, or from equations like $F_k = \mu_k N$. Sometimes you are short one (it should be obvious from simple counting); then you have to think about the special features of the problem—for example, two tensions have to be equal, two distances have to add up to a third from simple geometry, or the normal force must be zero (as in Problem IV). The precise form of these equations depends on the variables you choose (try redoing Problem IV in terms of an angle instead of r and h), but any way you do it is OK. By now you should have picked up a few tricks of the trade—you shouldn't have to draw a little triangle, for example, every time you have to distinguish between $\cos\theta$ and $\sin\theta$; you should be able to decide very easily when torques are positive and when they are negative. When you finally have enough equations and are convinced they're correct, you should congratulate yourself—you're done with the physics and you're just left with math!

Step 3: Solve the Equations. After doing the hairy trigonometry of Problems V and VI, you'll probably agree that this is often easier said than done. But at least you've separated problems in doing the math from problems in understanding the physics. If you think that you've made a computational error somewhere along the line, use dimensional analysis on your equations. Unfortunately, dimensional analysis doesn't detect sign errors; to get rid of these you'll have to think which signs in your equations don't make sense physically.

Step 4: Plug In the Numbers to Get an Answer. It's true that you force yourself to use algebra instead of arithmetic if you wait until the very end to plug in. But it's well worth it: it's much easier to find and correct errors in algebra than errors in arithmetic; also, you get much more as a final answer. Think of Problem VI. The numerical answer was 63.4 cm. You can't do much with this answer; you can't even check whether it's right. The formula for the answer is

$$x_{\text{CM}} = \frac{L}{[1 + (\tan\theta/\tan\phi)]} .$$

You can be pretty confident that this is right; it's dimensionally correct and works in the three special cases of ϕ near $90°$, θ near $90°$, and $\phi = \theta$.

ANSWER KEY

1. 10 lb
2. $T_1 = T_2 = (L - l - l_1 - l_2)\left(\dfrac{k_1 k_2}{k_1 + k_2}\right)$
3. 56.7 N; 19.5° above the horizontal
4. One convenient pivot point is at the block. Then $F_1 r_1 = F_2 r_2$. However, you can eliminate an unknown immediately by choosing the pivot point at the top of pillar 1 or the top of pillar 2. Then you don't have to use the condition from translational equilibrium! Try all three ways to verify that they all give the same answer.
5. No
6. Where the bump meets the wheel. This eliminates the force exerted by the bump.
7. $\phi = \tan^{-1}\left[\dfrac{(m_1 - m_2)}{(m_1 + m_2)} \cot\theta\right]$
8. $T_1 \sin\theta + T_2 \sin\phi = mg$ (no vertical acceleration); $T_1 \cos\theta = T_2 \cos\phi$ (no horizontal acceleration)
9. Let F increase from 0 in your mind. At a certain value F_c, the normal force on the wheel from the ground (not the bump) will be 0. For F just slightly greater than F_c, the wheel will move over the bump. So you must find F_c for the static situation with no normal force at the bottom of the bump.
10. $\sum F_x = \sum F_y = 0$
11. 10 lb
12. Sum torques about the lower end of the plank.
13. No
14. $\mu_s = \dfrac{l \sin^2\theta \cos\theta}{2h - l \cos^2\theta \sin\theta}$
15. $F = \dfrac{mgl}{2h}\cos\theta \sin\theta$, perpendicular to the plank.
16. $\mu_s = \dfrac{H}{V}$
17. $\cos(\theta \pm \phi) = \cos\theta \cos\phi \mp \sin\theta \sin\phi$.
18. $F(r - h) - W\sqrt{r^2 - (r-h)^2} = 0$.
19. $F_1 = 100$ N, up; $F_2 = 200$ N, up
20. $T_1 + T_2 + T_3 + T_4 = mg$
    ($\mathbf{F} = m\mathbf{a}$, vertically);
    $(T_1 + T_2)x_{CG} = (T_3 + T_4)(x - x_{CG})$
    $(T_1 + T_4)z_{CG} = (T_2 + T_3)(z - z_{CG})$
    (from torques);
    $T_1 = ky_1,\ T_2 = ky_2,$
    $T_3 = ky_3,\ T_4 = ky_4;$
    $(y_1 + y_3)/2 = (y_2 + y_4)/2$
    (midpoints of diagonal lines connecting tops of springs have the same height!)
21. They're each $r\cos\phi$; their ratio is independent of ϕ. Thus the balance has no equilibrium position; once set in motion, it keeps tipping until a tray hits the table.
22. $x_{CM} = \dfrac{L}{[1 + (\tan\theta/\tan\phi)]}$
    $\simeq 63.4$ cm
    to the right of the left end point.
23. Three
24. $F_1 = 100$ N, down; $F_2 = 400$ N, up
25. $F_{\min} = W\dfrac{\sqrt{h(2r - h)}}{r - h}$.
    Does this expression make sense as $h \to r$, $h \to 0$, $W \to 0$, ...?

26. $l_1 + x_1 + l + x_2 + l_2 = L$
27. Two unknowns, one equation
28. $r\cos(\theta + \phi)$; $r\cos(\theta - \phi)$
29. A frictionless roller cannot support a tangential force.
30. $F_1 + F_2 - mg = 0$
31. 10 lb
32. $\theta = \phi$
33. $mgx = T_2 L \sin\phi$

learning guide 7

Gravitation

Suggested Reading: Fishbane, Gasiorowicz, & Thornton, Chapter 12

PROBLEM I

1. The moon circles the earth with period $T = 27.3$ days. The distance from the center of the earth to the center of the moon is $R = 3.85 \times 10^5$ km. What is G, the gravitational constant, in terms of T, R, and M, the mass of the earth? If you're having trouble, use Helping Questions 1 and 2.
    Key 32
2. What is G in terms of M only? Use SI units.   **Key 28**
3. An apple falls to the earth with acceleration $g = 9.8$ m/s^2. The radius of the earth is $r = 6370$ km. What is G in terms of g, r, and M? For assistance, see Helping Question 3.   **Key 15**
4. What is G in terms of M only? Use SI units.   **Key 26**

PROBLEM II

1. Is it possible for an orbiting satellite always to remain above the same point on the earth? If so, how many kilometers above the earth's surface must

it be? Use the value for GM that you calculated in Problem I. Look at Helping Questions 4 and 5 if you need them. **Key 12**

PROBLEM III

A distant star, observed by telescope over the course of many years, is found not to remain fixed but rather to execute uniform circular motion for no apparent reason. It is hypothesized that the cause of the star's motion is a *black hole*. One possibility is sketched in the diagram. The star and the black hole move on the same circular path.

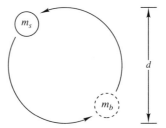

1. What can you say about the mass of the star m_s and the mass of the black hole m_b? Are you unable to see any relation? Look at Helping Question 6. **Key 24**
2. The period of the star's motion is T. What is the mass of the black hole in terms of g, T, and d? If need be, see Helping Questions 7 and 8. **Key 13**

PROBLEM IV

A meteor moves toward the solar system with speed v_0 in a direction such that it would miss the sun by a distance d if it were not attracted by the sun's gravitational force. Denote the mass of the sun by M.

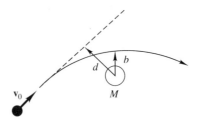

1. Find the distance b of the meteor from the sun at the point of closest approach in terms of v_0, d, and M (and G, the gravitational constant). Turn to Helping Questions 9 to 11 if you're stumped. **Key 9**

Learning Guide 7 Gravitation

PROBLEM V

Three particles all of mass m are located at the vertices of an equilateral triangle and are spinning about their center of mass in otherwise empty space. The sides of the triangle are of length d, which does not change with time.

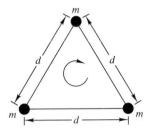

1. What is the potential energy of the system? If you don't see how to get the answer, reread Fishbane, Gasiorowicz, & Thornton, Section 12-3. **Key 2**
2. What is the kinetic energy of the system in terms of G, m, and d? Puzzled? See Helping Questions 12, 13, and 14. **Key 4**

PROBLEM VI

In Section 12-3, Fishbane, Gasiorowicz, & Thornton prove Kepler's third law for circular orbits. In this problem you will prove it for the more general case of elliptical orbits.

A planet of mass m orbits a star of mass M. M is much larger than m, so the star can be considered stationary. The orbit is an ellipse and the star is at one of its foci. The total energy of the system is negative, so for convenience call it $-E$ where E is positive. Let the magnitude of the planet's angular momentum be l.

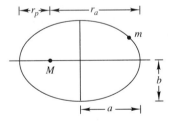

1. In terms of E, l, m, M, and G, find r_p and r_a (called the perihelion and aphelion radii of the orbit). If you have difficulty, try Helping Questions 15, 16, and 17, one by one. **Key 21**

For every point on an ellipse, the sum of the distances from the point to the two foci is the same.

2. Use this fact to show that

$$a = \frac{r_a + r_p}{2}$$
$$b = \sqrt{r_a r_p}$$

If you have trouble showing the second equation, use Helping Question 18.

3. Express a and b in terms of E, l, m, M, and G. **Key 3**

Kepler's second law says that the rate at which area is "swept out" by the planet's orbit is $l/2m$.

4. Use the second law to find the period of the orbit in terms of G, M, and a only. Stuck? Use Helping Questions 19 and 20. **Key 25**

HELPING QUESTIONS

1. Can you combine Newton's second law and Newton's law of gravitation into one equation that relates the moon's acceleration a to M, G, and R? **Key 14**

2. What is the moon's acceleration a in terms of R and T? **Key 1**

3. Can you combine Newton's second law and Newton's law of gravitation to get an equation relating the given variables? **Key 22**

4. What would be the orbital period of such a satellite? **Key 16**

5. What is the orbital radius for a satellite circling the earth with period T? **Key 23**

6. If the masses of the stars were different, would the center of mass be fixed throughout the motion drawn in the diagram? **Key 17**

7. What is the force on the black hole? **Key 29**

8. What is the acceleration of the black hole? **Key 30**

9. What *two* physical quantities are conserved? **Key 10**

10. What is the total energy of the system when the meteor is far away? At its closest point? Use m for the mass of the meteorite and v for the speed of the meteorite at its point of closest approach. **Key 18**

11. What is the angular momentum of the meteor about the sun when it is far away? When it's at its closest point? **Key 5**

12. What is the kinetic energy of the system in terms of m, d, and ω? **Key 20**

13. What is the acceleration a of each mass in terms of ω and d? **Key 8**

14. What is the net force on each mass in terms of G, m, and d? **Key 31**

15. At any given time, let r be the distance to the planet from the sun, let v_r be the radial component of the velocity, and let $v_\perp$ be the component of the velocity which is perpendicular to the radius vector. In terms of these parameters and

others, write an expression for
E and an expression for *l*.
Key 11

16. At perihelion and aphelion, what is the value of v_r? **Key 33**

17. Taking for v_r the value just found, the two equations you wrote down in Helping Question 15 can be solved simultaneously for *r*. How many solutions are there? What do the solutions mean? **Key 7**

18. What are the lengths *c* and *d* in terms of r_a and r_p? Now use the Pythagorean theorem. **Key 19**

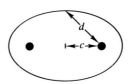

19. What is the orbital period in terms of the area *A* of the ellipse, *l*, and *m*? **Key 6**

20. What is the area of the ellipse in terms of *a* and *b*? **Key 27**

Notes: Conserved Quantities Again

The notes for Learning Guide 3 said that if you can solve a problem by conservation principles, it is probably easier that way than by $\mathbf{F} = m\mathbf{a}$. On the other hand, for some problems the conservation laws by themselves simply don't give you enough information to solve the problem. Then you have to use $\mathbf{F} = m\mathbf{a}$. Let's look at three examples from this Learning Guide.

The only way to solve Problem IV is by using both conservation of energy and conservation of angular momentum. If you wanted, you could write down the differential equations that you get from the components of $\mathbf{F} = m\mathbf{a}$. They're just

$$\frac{d^2x}{dt^2} = GMm \frac{x}{(x^2 + y^2)^{3/2}}$$

$$\frac{d^2y}{dt^2} = GMm \frac{y}{(x^2 + y^2)^{3/2}}.$$

But these equations would require advanced techniques to solve.

Notice that slight changes to the problem put the solution out of reach of conservation principles. Suppose the problem asked how much *time* it would take for the meteor to go from its initial point to its point of closest approach. Then you couldn't do the problem using only conservation laws; since time doesn't appear in the equations that the conservation laws give you, you'd have to use the *differential equations* written above. Or suppose the problem asked for the *angle* that the meteor's final velocity makes with its initial velocity. Once again, you'd have to use the differential equations written above.

Problem V is another instructive case. You solved this problem by using forces. However, the only reason you were able to avoid using a differential equation was that the motion was simple. If the particles had the initial velocities shown in the diagram, there is no way you'd be able to calculate how the kinetic energy and the potential energy vary with time. You could write down the differential equation governing the motion, but you wouldn't be able to solve it. (No one has ever solved this differential equation.)

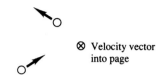

When you get to thermodynamics you'll see more applications of conservation laws.

ANSWER KEY

1. $a = \left(\dfrac{2\pi}{T}\right)^2 R$

2. $-\left(\dfrac{Gm_1m_2}{d} + \dfrac{Gm_2m_3}{d} + \dfrac{Gm_1m_3}{d}\right) = -3\dfrac{Gm^2}{d}$

3. $a = \dfrac{GMm}{2E}$; $b = \dfrac{l}{\sqrt{2Em}}$

4. $3Gm^2/2d$

5. $mv_0 d$; mvb

6. $T = 2mA/l$

7. Two—the smaller one is r_p and the larger one is r_a.

8. $a = \dfrac{\omega^2 d}{\sqrt{3}}$

9. $b = \dfrac{-GM + \sqrt{(GM)^2 + v_0^4 d^2}}{v_0^2}$

10. Total energy and angular momentum about the sun

11. $E = \dfrac{GMm}{r} - \dfrac{1}{2}m(v_\perp^2 + v_r^2)$; $l = mv_\perp r$

12. Yes: $\sqrt[3]{\dfrac{GMT^2}{4\pi^2}}$ (radius of the earth) $- R_e \simeq 36000$ km above a point on the equator.

13. $m_b = \dfrac{2\pi^2 d^3}{T^2 G}$

14. $a = \dfrac{GM}{R^2}$

15. $G = gr^2/M$

16. 24 h

17. No

18. $\dfrac{1}{2}mv_0^2$; $\dfrac{1}{2}mv^2 - \dfrac{GMm}{b}$

19. $c = \dfrac{r_a - r_p}{2}$; $d = \dfrac{r_a + r_p}{2}$

20. $K = \dfrac{1}{2}(3m)\omega^2\left(\dfrac{d}{\sqrt{3}}\right)^2 = \dfrac{1}{2}m\omega^2 d^2$

21. $r_p = \dfrac{GMm - \sqrt{(GMm)^2 - 2El^2/m}}{2E}$
    $r_a = \dfrac{GMm + \sqrt{(GMm)^2 - 2El^2/m}}{2E}$

22. $g = GM/r^2$

Learning Guide 7   Gravitation

23. $\sqrt[3]{\dfrac{GMT^2}{4\pi^2}}$

24. $m_s = m_b$

25. $\dfrac{2\pi}{\sqrt{GM}} a^{3/2}$

26. $\dfrac{4.0 \times 10^{14}}{M} \dfrac{\text{m}^3}{\text{s}^2}$

27. $A = \pi a b$

28. $(4.0 \times 10^{14}\ \text{m}^3/\text{s}^2)/M$

29. $\dfrac{G m_s m_b}{d^2}$

30. $a = \dfrac{d}{2}\left(\dfrac{2\pi}{T}\right)^2$

31. $F = \dfrac{\sqrt{3} G m^2}{d^2}$

32. $G = \dfrac{4\pi^2 R^3}{T^2 M}$

33. $v_r = 0$

learning guide 8

Simple Harmonic Motion

Suggested Reading: Fishbane, Gasiorowicz, & Thornton, Chapter 13

PROBLEM I

An ideal spring of spring constant k is attached to a stationary wall on one end and a block of mass m on the other. The mass is pulled a distance A to the right of its equilibrium position and released from rest. Assume there is no friction between the table and the block.

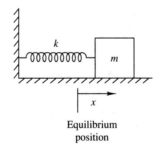

Equilibrium position

1. What is the angular frequency ω of the system in terms of k and m? If you don't know, take another look at Fishbane, Gasiorowicz, & Thornton. Otherwise, use the text as little as possible for the rest of this problem.

Key 26

2. What is the position of the block $x(t)$ as a function of time? Express your answer in terms of A, ω, and t. Take the zero of the x-axis to be at the equilibrium point of the spring and the positive direction toward the right.
Key 5

3. What is the block's velocity $v(t)$? Express your answer in terms of A, ω, and t.
Key 31

4. What is the system's potential energy $U(t)$? Write your answer in terms of A, k, ω, and t.
Key 10

5. What is the block's kinetic energy $K(t)$? Express your answer in terms of A, m, ω, and t.
Key 21

6. Verify that the total energy $E(t)$ of this system is constant.
Key 16

PROBLEM II

A system like the one of Problem I is at rest at equilibrium. A bullet of mass m moving with velocity v embeds itself in the block. Assume that the collision between the bullet and the block takes place in such a short time interval that during the collision the spring compresses by a negligible amount. The block will, however, be set in motion by the bullet's impact.

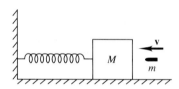

1. What is the amplitude A of the resulting motion? Use Helping Questions 1 and 2.
Key 33

PROBLEM III

An object is hung by a wire from the ceiling. The object has rotational inertia I about the axis made by the extended wire. When the object is twisted through an angle θ, the wire exerts a torque τ that tends to restore the system to equilibrium. The scalar dependence of τ on θ is Hooke's law $\tau = -k\theta$, where the minus sign indicates the restoring nature of the torque and k is a constant for the wire (analagous to k for a linear spring).

Learning Guide 8   Simple Harmonic Motion

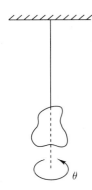

1. Write down a differential equation that the motion $\theta(t)$ must obey. The text has a complete solution of this problem, but Helping Question 3 should be enough if you're confused.   **Key 1**
2. Write down the general solution to this equation. For assistance, see Helping Question 4.   **Key 6**
3. If the wire is shortened by a factor of 2, will the period be larger or smaller? See Helping Questions 5 and 6.   **Key 22**

PROBLEM IV

The pendulum drawn in the diagram swings without friction under the influence of gravity. Assume that the rod holding the body is massless and rigid and doesn't stretch.

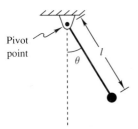

1. Write down a differential equation that the motion $\theta(t)$ must satisfy exactly. See Helping Question 7   **Key 42**
2. Approximate this differential equation for the case of small amplitudes by a differential equation you can solve. Solve it. Use Helping Question 8 if you can't think of a good approximation.   **Key 23**
3. If the pendulum's length is made shorter, does its period become larger or smaller?   **Key 9**

PROBLEM V

In the situation shown in the sketch, the springs are ideal and the surfaces frictionless. Both springs have the same equilibrium extension. The mass moves in simple harmonic motion with angular frequency ω. Suppose that the two springs are replaced by a single spring with spring constant k and the mass moves with the same angular frequency ω.

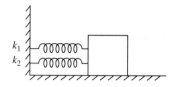

1. What must k be in terms of k_1 and k_2? Use Helping Question 9 if you're lost. **Key 18**

In a second situation, the same two springs are put into the second configuration shown. The mass moves with angular frequency ω'. Again suppose that the two springs are replaced by a single spring with spring constant k' and the mass moves with the same angular frequency ω'.

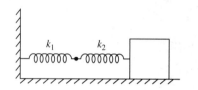

2. What must k' be in terms of k_1 and k_2? Use Helping Questions 10 and 11 if you need to. **Key 39**

Thus with two springs, you can make four "effective springs," with spring constants k_1 (the first spring alone), k_2 (the second spring alone), k (the springs "in parallel"), and k' (the springs "in series").

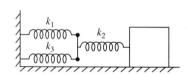

3. Just for fun, how many effective springs can you make with three springs? You are allowed to have series, parallel, and series-parallel (like the one in the sketch) combinations. Assume that $k_1, k_2,$ and k_3 are chosen so that there are no "accidental equivalences" such as

$$k_1 = k_2 + k_3$$

making

$$\underset{k_1}{\text{000000000}} \equiv \begin{bmatrix} \underset{k_2}{\text{00000000}} \\ \underset{k_3}{\text{00000000}} \end{bmatrix}$$

(**Hint:** Draw pictures!) **Key 44**

Learning Guide 8   Simple Harmonic Motion

PROBLEM VI

The curve in the sketch is the path of a particle executing simple harmonic motion both horizontally and vertically; its path is given by the equations of motion

$$x(t) = A_x \cos(\omega_x t + \phi_x)$$
$$y(t) = A_y \cos(\omega_y t + \phi_y).$$

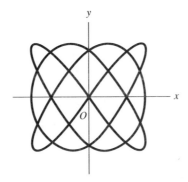

1. For the motion shown in the diagram, what is the ratio of the periods T_x/T_y? See Helping Question 12 if you don't know how to get started.   **Key 29**

   The motion shown in the picture is periodic: the particle returns to its initial position with its initial velocity after a certain time T and then simply repeats its motion forever.

2. For what values T_x/T_y is the motion periodic? Need help? See Helping Question 13.   **Key 13**

PROBLEM VII

A block of mass m hangs from a spring of spring constant k. The block is pulled down a distance A from equilibrium and released from rest. The damping constant of the oscillator is b, and so the block moves according to

$$x = A e^{-bt/2m} \cos(\omega' t),$$

where ω' is a number slightly less than $\sqrt{k/m}$.

1. When will the amplitude be $A/2$? Use Helping Question 14 if you're stuck.   **Key 12**
2. Roughly what fraction of the original energy will have been lost by this time?   **Key 3**

PROBLEM VIII

A malicious man wants to destroy a strong spring by stretching it beyond its elastic limit. The spring is one meter long at equilibrium and has an elastic limit of 0.5 m. If it is stretched past 1.5 m or compressed past 0.5 m it will not regain its original shape and will be ruined. In between these extremes, it is an ideal spring of spring constant 10,000 N/m. The man has a mass of 100 kg.

1. The man tries to destroy the spring by hanging from it. Does he succeed?
   Key 4

The man drops back to the ground to look for paper and pencil. Next he plans to destroy the spring by delivering a force to the system that consists of the spring and his body. He plans to vary the force sinusoidally in time at precisely the resonant frequency of the spring-body system. He writes down the following equation (which is correct and graphed in Figure 13-21 of the text):

$$A = \frac{F_m}{\sqrt{m^2(\omega_e^2 - \omega^2)^2 + b^2\omega_e^2}}.$$

In this equation, F_m is the maximum value of the sinusoidally varying driving force, ω_e is the angular frequency of the driving force, ω is the natural angular frequency of the spring-body system, and b is the damping constant of the system. A is the amplitude of the resulting motion.

2. He estimates the damping constant at 1 kg/s. He plans to set ω_e equal to ω. How many cycles per second is this, and how many newtons must F_m be so that $A = 0.5$ m? **Key 15**

The man jumps back on the spring to put his idea into practice. He simulates the external force by slightly oscillating his body. But he can't quite oscillate with a frequency $\nu = 1.6$ Hz. He does manage to oscillate with $\nu = 1$ Hz.

3. Calculate roughly what F_m has to be for A to be 0.5 m at this lower driving frequency. Does he succeed in destroying the spring? **Key 20**

PROBLEM IX

In this problem you will examine an example of *coupled oscillations*. Consider two blocks of equal mass and three identical springs arranged as shown in the figure. Ignore friction throughout this problem.

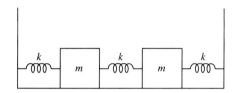

1. If x_1 and x_2 represent the displacement from the equilibrium position of the respective blocks, show that

$$m\frac{d^2 x_1}{dt^2} = k(x_2 - 2x_1) \quad \text{and} \quad m\frac{d^2 x_2}{dt^2} = k(x_1 - 2x_2).$$

Key 14

2. Find all the values of ω for which there are values of A_1 and A_2 such that

$$x_1(t) = A_1 \sin(\omega t)$$
$$x_2(t) = A_2 \sin(\omega t)$$

is a solution to the above pair of differential equations. If you can't find a good method of attack, see Helping Question 15. **Key 30**

What is the physical meaning of the two angular frequencies you found in part (2)? In parts (3) and (4) you will discover the answer to this question.

3. Imagine that the middle spring is replaced by a massless, inextensible rod so that the two masses oscillate back and forth together ($A_1 = A_2$). What is their angular frequency of oscillation? Use Helping Question 16 if your answer doesn't check. **Key 8**

4. Now imagine that the middle spring is back in place but held fixed at its center. The masses oscillate symmetrically toward and away from each other ($A_1 = -A_2$). What is their angular frequency of oscillation? Use Helping Question 17 if you need help. **Key 43**

If you take a linear algebra course you will learn a more systematic approach to part (2) that generalizes to more springs and more blocks with different spring constants and masses. In the language of linear algebra, the heart of the method is finding the *eigenvalues* and *eigenvectors* of a *symmetric matrix*.

HELPING QUESTIONS

1. Is momentum conserved during the collision? Then what is the velocity $v_{\max}$ of the block immediately after the collision? **Key 17**

2. Is energy conserved after the collision? Can you get an equation relating the amplitude of the resulting motion to the block's velocity just after the collision? **Key 11**

3. What is Newton's law for rotational motion? Can you combine it with Hooke's law to get a differential equation for θ? **Key 35**

4. You know that the general solution of
$$m\frac{d^2x}{dt^2} = -kx$$
is
$$x(t) = A\,\cos\left(\sqrt{\frac{k}{m}}t + \phi\right).$$
Does this help? **Key 28**

5. What is the torsional pendulum's period T in terms of κ and I? **Key 38**

6. Use your physical intuition to decide whether κ would increase or decrease when the wire is shortened. **Key 25**

7. Draw a force (free-body) diagram for the mass. What is the net tangential force? Now use $\tau = I\alpha$. **Key 27**

8. For small θ, what is $\sin\theta$ approximately equal to? **Key 37**

9. In each case, if the mass is displaced a distance x, what is the force on the mass? **Key 41**

10. If the force on the mass is F, what are the magnitudes of the two forces acting on spring 2? The two forces acting on spring 1? **Key 32**

11. If the force on the mass is F, what is the extension x_1 of spring 1? The extension x_2 of spring 2? Can you express $F/(x_1 + x_2)$ in terms of k_1 and k_2? **Key 24**

12. Pick an arbitrary point along the path to serve as your starting point. How many cycles n_x in the x-direction does the particle make before returning? How many cycles n_y in the y-direction? Suppose the total time around the path is T. Express T in terms of n_x and T_x, and also in terms of n_y and T_y. **Key 2**

13. What conditions on ω_x and ω_y make the motion periodic with period T? Find an expression for T_x/T_y in terms of n_x and n_y. **Key 19**

14. Which factor of the equation $x = Ae^{bt/2m}\cos(\omega' t)$ has to do with the decrease in amplitude? What is the value of this factor when the amplitude is $A/2$? **Key 7**

15. Plug in the conjectured solution, differentiate, and eliminate the time t to yield two algebraic equations. Eliminate ω temporarily to get a quadratic equation relating A_1 to A_2. What is this quadratic equation? Use the relations between A_1 and A_2 that this quadratic equation gives you to find the possible values of ω. **Key 34**

16. Use what you learned in Problem V about combining springs. What is the effective spring constant? What is the total mass? **Key 36**

17. If a spring is cut in half, what happens to its spring constant? Now use what you know about combining springs. **Key 40**

Learning Guide 8   Simple Harmonic Motion

Notes: Differential Equations

You have probably noticed that this Learning Guide has a slightly different character from the first seven Learning Guides. There is a reason for this—the subject matter has changed slightly. *You have now learned the principles of classical mechanics, and are moving on to its applications!*

Earlier on you studied the basic principles of mechanics—Newton's second law ($\mathbf{F} = m\mathbf{a}$ for a single particle) and Newton's third law. As time went on you studied some direct implications of these two laws—conservation of energy, Newton's second law for systems of particles ($\mathbf{F}_{\text{ext}} = M\mathbf{a}_{\text{CM}}$), conservation of momentum, and Newton's second law for rotational motion ($\boldsymbol{\tau} = d\mathbf{L}/dt$). In the coming Learning Guides you will study applications of these principles—simple harmonic motion, fluid mechanics, and waves. As you've learned in this Learning Guide, applying Newton's laws is easier said than done. The heart of the difficulty is that differential equations occur naturally, and some are hard to solve. How do differential equations get into physics? Because Newton's second law itself is a differential equation:

$$F = ma = m\frac{d^2x}{dt^2}.$$

Let's compare solutions to this differential equation for three different force laws:

Constant force F	Hooke's law $F = -kx$	Pendulum $F = mg\sin\theta$
$\frac{d^2x}{dt^2} = \frac{F}{m}$	$\frac{d^2x}{dt^2} = -\frac{k}{m}x$	$\frac{d^2\theta}{dt^2} = -(g/\ell)\sin\theta$
↓	↓	↓
$\frac{dx}{dt} = v_0 + \frac{F}{m}t$		
↓	↓	↓
$x = x_0 + v_0 t + \frac{1}{2}\frac{F}{m}t^2$	$x = A\sin\left(\sqrt{\frac{k}{m}}t + \phi\right)$	? ? ? ?

When you first learned $\mathbf{F} = m\mathbf{a}$, the problems you were asked to do involved constant forces. Although you may not have realized it at the time, you already knew the solution to the resulting equation of motion—i.e., you knew that for constant acceleration in the x direction the resulting displacement is

$$x(t) = x_0 + v_{0x}t + \tfrac{1}{2}at^2.$$

Hooke's law, however, leads to a more difficult differential equation to solve; in effect, Fishbane, Gasiorowicz, & Thornton "solved" it by just guessing the right solution! To see that their guess is indeed correct, you need only plug it back into the original differential equation. Had their guess been wrong, it would have led to a contradiction. The guess is a perfectly acceptable and time-honored method of solving differential equations, but it has an obvious limitation. If you take a course in differential equations, you will learn a more systematic way to solve this differential equation and also the differential equations of damped and forced oscillations,

$$m\frac{d^2x}{dt^2} + kx + b\frac{dx}{dt} = 0 \quad \text{and} \quad m\frac{d^2x}{dt^2} + kx + b\frac{dx}{dt} = F_m \cos(\omega_e t).$$

Most differential equations in physics are more like the pendulum equation—they are either very difficult or actually impossible to solve exactly. So techniques for finding approximate solutions are extremely important.

Here are some questions to think about relating to approximations in three of the problems you've just finished. The answers are at the end of the notes.

Problem IV. How well does the approximate solution to the pendulum equation

$$\theta(t) = A \sin\left(\sqrt{\frac{g}{\ell}}t + \phi\right)$$

describe the real solution? Are the two periods similar? For the same initial conditions, would $\theta_{\text{approx}}(t) \simeq \theta_{\text{real}}(t)$ for large values of t?

Problem VI. If the motion had been given by the differential equations

$$\frac{d^2x}{dt^2} = -k_1 \sin x \quad \text{and} \quad \frac{d^2y}{dt^2} = -k_2 \sin y,$$

could you have used methods of approximation to decide when the motion was periodic?

Problem VIII. Do you think the force of friction is well approximated by $-b\,dx/dt$ in the diagram? When you study electricity and magnetism you will become an expert in turning impossible problems into solvable problems by making small approximations.

One thing that decreases the difficulty caused by differential equations is that, often in physics, the same differential equation occurs in very different physical contexts. In particular, the differential equation for simple harmonic motion applies very often, especially if one is only interested in the approximate behavior of a system. As you saw in Problems III and IV, differential equations are no problem if you already know how to solve them! As you study more physics you will spend more time learning how to solve and intuitively understand the differential equation for forced oscillations in a variety of contexts. For example, you will see that this equation also applies to electrical oscillations.

Answers to the Preceding Questions

IV: For small oscillations, the periods are similar but $\theta_{\text{approx}}(t)$ does not approximate $\theta_{\text{real}}(t)$ for large values of t (the two will be on different sides of equilibrium half the time).

VI: No.

VIII: It would be a poor approximation. As you know, a better approximation for the force of friction is $\mu_k N$—the frictional force depends only on the direction of the velocity, not its magnitude.

ANSWER KEY

1. $I\dfrac{d^2\theta}{dt^2} = -\kappa\theta$

2. Three; four; $T = n_x T_x = n_y T_y$

3. About $\tfrac{3}{4}$. The energy is not exactly $k/2$ times the amplitude squared; it is sometimes above this slightly and sometimes below. You can see this by looking at the power $\mathbf{F}\cdot\mathbf{v}$ lost to friction.

4. No. He just stretches it 10 cm.

5. $x(t) = A\cos(\omega t)$

6. $\theta(t) = A\cos\left(\sqrt{\dfrac{\kappa}{I}}t + \phi\right)$

7. $e^{-bt/2m}$; it equals $\tfrac{1}{2}$.

8. $\omega = \sqrt{k/m}$

9. Period becomes smaller.

10. $U(t) = \tfrac{1}{2}kA^2\cos^2(\omega t)$

11. Yes;
$\tfrac{1}{2}(M+m)v_{\max}^2 = \tfrac{1}{2}kA^2$

12. $t = \dfrac{2m\ln 2}{b}$

13. The motion is periodic when T_x/T_y is a rational number. It is not periodic when T_x/T_y is an irrational number.

14. Draw a force (free-body) diagram of each spring and block. Then use Newton's law and Hooke's law.

15. About 1.6 Hz; 5 N

16. $E(t) = U(t) + K(t)$
$= \tfrac{1}{2}kA^2\cos^2(\omega t)$
$+ \tfrac{1}{2}mA^2\omega^2\sin^2(\omega t)$
$= \tfrac{1}{2}kA^2\cos^2(\omega t)$
$+ \tfrac{1}{2}kA^2\sin^2(\omega t)$
$= \tfrac{1}{2}kA^2$

17. Yes—since the spring compresses only negligibly during the collision, there are essentially no external forces on the block-bullet system, and so
$$v_{\max} = \dfrac{m}{M+m}v$$

18. $k' = k_1 + k_2$

19. $\omega_x T = 2\pi n_x$ and $\omega_y T = 2\pi n_y$, where n_x and n_y are *integers*. $T_x/T_y = n_y/n_x$

20. F_m would have to be about 3000 N, so he doesn't succeed.

21. $K(t) = \dfrac{1}{2}mA^2\omega^2\sin^2(\omega t)$

22. Smaller

23. $\ell\dfrac{d^2\theta}{dt^2} \simeq -g\theta$
$\theta = A\cos\left(\sqrt{\dfrac{g}{\ell}}t + \phi\right)$

24. F/k_1; F/k_2; yes: after you plug in for x_1 and x_2, the F's cancel.

25. Increase. For a given θ, the wire is twisted more *per unit length*. Thus the torque it exerts on the object is increased.

26. $\omega = \sqrt{k/m}$

27.

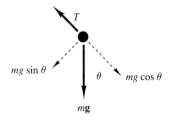

Net tangential force $= mg\sin\theta$.

28. It helps—the differential equation for the motion of a mass on a spring is the same as the differential equation for the motion of the torsional pendulum, except that the names of the variables are changed.

29. 4/3
30. $\sqrt{k/m}$; $\sqrt{3k/m}$
31. $v(t) = -A\omega \sin(\omega t)$
32. All four forces have magnitude F.
33. $A = v\sqrt{\dfrac{m^2}{(M+m)k}}$
34. $A_1^2 = A_2^2$, so $A_1 = \pm A_2$.
35. $\tau = I\alpha = I\dfrac{d^2\theta}{dt^2}$.
    Yes—you now have two expressions for torque, so you can set them equal.
36. $2k$; $2m$
37. θ
38. $T = 2\pi\sqrt{\dfrac{I}{\kappa}}$
39. $k' = \dfrac{k_1 k_2}{k_1 + k_2}$
40. It doubles.
41. $-k_1 x - k_2 x$; $-kx$
42. $\ell\dfrac{d^2\theta}{dt^2} = -g\sin\theta$
43. $\omega = \sqrt{3k/m}$
44. 14 combinations + 3 original springs

learning guide 9

Waves

Suggested Reading: Fishbane, Gasiorowicz, & Thornton, Chapters 14 and 15

PROBLEM I

A wave travels on a hypothetical stretched string of infinite length. At time t, the element of the string at position x is displaced a distance

$$y = A \sin(x - t),$$

where $A = 1$ cm $= 0.01$ m, x has units of meters, and t has units of seconds. The graph shows the shape of the string at a fixed point in time (the scales of the two axes are different).

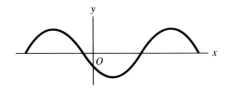

1. What is the angular frequency ω? The frequency f? If you have trouble with this part or parts (2) through (5), reread Fishbane, Gasiorowicz, & Thornton, Section 14-3. **Key 9**

2. What is the wave number k? **Key 32**
3. What is the period T? **Key 10**
4. What is the wavelength λ? **Key 28**
5. What is the phase velocity v? **Key 37**
6. What is the amplitude y_m? **Key 30**
7. Is energy transferred to the left or to the right? If you can't answer this or the next part, reread Section 14-5. **Key 2**
8. What else do you need to know before you can calculate the amount of energy transferred per unit time? Does it make sense physically that two waves of the same size, shape, and speed can transfer different amounts of energy? **Key 16**

PROBLEM II

Let the mass per unit length of the string in Problem I be μ.

The apparatus in the sketch is a collection of an infinite number of simple harmonic oscillators. The mass hanging from each spring is a segment of length $\Delta \ell$ of the string of Problem I. The spring constants and the initial positions and velocities are chosen so that the motion of the string is precisely what it was in Problem I.

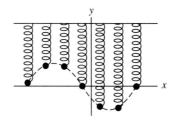

1. What is the total energy contained in the oscillators per unit length of the string? Express your answer in terms of y_m, the frequency f, and μ. Turn to Helping Questions 1 and 2 if you're stuck. **Key 8**
2. One could define the "average rate of energy transfer" for the system of oscillators to be $P = Ev$, where E is the energy per unit of length calculated in part (1) and v is the phase velocity of the wave. But is energy really being transferred horizontally in the collection of oscillators? Was it really being transferred in the string of Problem I? **Key 27**

PROBLEM III

The wave function of a guitar string vibrating without any overtones is

$$y(x, t) = 0.003 \times \sin(4x) \times \cos(2080\, t) \text{ m},$$

where x and y are in meters and t is in seconds.

1. Verify that this is a reasonable wave function by calculating the length of the string and the pitch of the tone produced. Use Helping Question 3 for the length of the string and Helping Question 4 for the pitch of the tone.
   Key 20
2. What are the phase velocities of the traveling waves that sum to the standing wave? See Helping Question 5 for a hint. **Key 34**
3. What is the maximum speed of the midpoint of the stretched string? Clueless? See Helping Question 6. **Key 23**

PROBLEM IV

Since the wave motion of a guitar string is damped, the wave function given in Problem III is unrealistic. Indeed, the very fact that mechanical energy from the string is transformed into sound energy tells you that the string's motion must be damped.

1. The guitar string of Problem III vibrates with an *initial* amplitude of 3 mm. The mass per unit length of the string is 1 g/m. Assume first that there is no loss of energy to heat. About how many dB is the sound intensity one meter in front of the guitar? You will have to make some approximations and some educated guesses, but you should be within 15 dB of the Answer Key. Use Helping Questions 7 through 10. **Key 5**
2. Look at Figure 14-17 in Fishbane, Gasiorowicz, & Thornton. Is your answer to part (1) reasonable? About what fraction of the mechanical energy is lost to heat? **Key 39**

PROBLEM V

A 37.5-cm-long metal pipe is closed at one end. A small loudspeaker is placed near the open end. The speed of sound in the pipe is 333 m/s.

1. At what frequencies will resonance occur in the pipe as the sound frequency emitted by the speaker is varied from 200 to 1200 Hz? If your answer doesn't check with the one in the key, take another look at Section 14-6 of Fishbane, Gasiorowicz, & Thornton. **Key 3**
2. At what frequencies would resonance occur if both ends were open? **Key 25**

PROBLEM VI

The ratio between the frequencies of any two adjacent keys on the piano is the same

$$\frac{f_{\text{higher}}}{f_{\text{lower}}} = 2^{1/12} = 1.06.$$

The lowest string on a piano has a frequency of 27 Hz, whereas the highest string on the piano has a frequency of 6500 Hz.

1. Using the criterion that beats can be detected by the ear up to frequencies of about seven per second, for which pairs of keys can you hear beats? See Helping Question 11. **Key 11**
2. Can you actually hear beats between the notes? **Key 17**

PROBLEM VII

A man is riding a train moving at 135 mi/h, which is one-fifth the speed of sound. As he approaches his friend who is standing by the tracks, he blows his trumpet with a frequency f.

1. What frequency does his friend hear? If you don't remember the Doppler equation, look again at Section 14-7 of Fishbane, Gasiorowicz, & Thornton. **Key 18**

On the return trip, the man's friend, still standing by the tracks, blows her trumpet with frequency f.

2. What frequency does the man in the train hear? **Key 1**
3. Why are your answers to parts (1) and (2) different—doesn't each person see the other approach at 135 mi/h? **Key 35**

PROBLEM VIII

The general Doppler equation derived in Section 14-7 of Fishbane, Gasiorowicz, & Thornton has to be modified to describe the Doppler effect for *light* waves exactly. However, even unmodified it is a good approximation.

A physicist runs a red light. He contests the ticket and explains to the judge that, because of the Doppler effect, the light looked green to him. The judge declares him innocent of running the red light but fines him instead one dollar for every mile per hour he was moving above 60 mi/h.

1. How many dollars is the physicist fined? The speed of light is about $c = 186{,}000$ mi/s, and the wavelengths of red and green light are about 6.3×10^{-7} m and 5.3×10^{-7} m, respectively. If you can't remember the formula, look at Section 14-7. **Key 36**

PROBLEM IX

(Optional. You might want to try this if you're comfortable with integrals. You don't have to be an expert by any means.)

A stretched wire has fundamental frequency $f = \omega/2\pi$. When it is plucked, a listener hears not only the fundamental frequency, but also higher frequencies. In this problem you will learn the mathematics necessary to answer the question, "Exactly which harmonics are present and how strong are they?" The secret of the answer lies in the superposition principle. Suppose the wire is pulled out near one of its ends as shown. As Fishbane, Gasiorowicz, & Thornton discuss, the function $f(x)$ can be decomposed in exactly one way into a sum of sine waves. That is, there is exactly one choice for the numbers $A_1, A_2, A_3, \ldots$ such that

$$f(x) = A_1 \sin x + A_2 \sin(2x) + A_3 \sin(3x) + \cdots$$

is true. Notice that once you know the *Fourier coefficients* $A_1, A_2, \ldots$ of the function $f(x)$, you know the strength of all the harmonics: the vertical displacement y of the wire at position x and time t is

$$y(x, t) = A_1 \sin x \cos(\omega t) + A_2 \sin(2x) \cos(2\omega t) + \cdots.$$

So the nth harmonic has amplitude A_n.

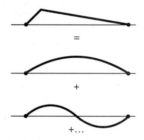

How do you get the coefficients $A_1, A_2, \ldots$? Before answering this question, let's take a detour that may seem completely irrelevant right now. Define the *dot product of two functions* $f(x)$ and $g(x)$ to be the number

$$f(x) \cdot g(x) \equiv \frac{2}{\pi} \int_0^\pi f(x) g(x) dx$$

If you're a calculus whiz, you can do the integrals to check that *the dot product of $\sin(n_1 x)$ and $\sin(n_2 x)$ equals 1 if $n_1 = n_2$ and 0 if $n_1 \neq n_2$. In other words, the functions $\sin x, \sin(2x), \ldots$ behave under their dot product the same as the unit vectors* **i**, **j**, *and* **k** *behave under theirs.*

1. Without doing the integral, show that $\sin x \cdot \sin(2x) = 0$. Use Helping Question 12.

The point of introducing the dot product for functions is that the nth Fourier coefficient A_n of a function $f(x)$ is just $f(x) \cdot \sin(nx)$. You can see this very easily, since

$$f(x) = A_1 \sin x + A_2 \sin(2x) + \cdots,$$
$$f(x) \cdot \sin(nx) = [A_1 \sin x + A_2 \sin(2x) + \cdots] \cdot \sin(nx)$$
$$= A_1(0) + A_2(0) + \cdots + A_n(1) + \cdots$$
$$= A_n.$$

This is completely analogous to the fact that you can get the components v_x, v_y, and v_z of a vector **v** by "dotting" **v** with **i**, **j**, and **k**.

The wire is plucked by pulling its center a distance y from equilibrium and then releasing the wire from rest.

2. Without doing any integrals, what can you say about the amplitude of the harmonics $2\omega, 4\omega, 6\omega, \ldots$? After a good effort, look at Helping Question 13. **Key 14**

If you know how to "integrate by parts," you can get the complete solution.

3. For every n, find the exact value of A_n. For hints, see Helping Questions 14 and 15. **Key 31**

HELPING QUESTIONS

1. How many oscillators are there per unit length? What is the maximum extension of each oscillator? **Key 29**

2. What is the spring constant of each oscillator? **Key 13**

3. What must ℓ be if $\sin(4x)$ equals zero at $x = 0$ and ℓ but never equals zero in between? **Key 24**

4. If $\omega = 2080\,\text{rad/s}$, what is the frequency f? Is this a typical musical note? **Key 22**

5. What are the wave functions of the two traveling waves? What is the phase velocity of a traveling wave of angular frequency ω and wave number k? **Key 12**

6. What is $\sin(4x)$ at the middle of the string? Then what is dy/dt? **Key 26**

7. From Problem III, the midpoint of the string has maximum speed about 6 m/s. So a "typical" point on the string would have a maximum speed of about 3 m/s. Roughly, what is the initial energy of the string? **Key 19**

8. Estimate how long it takes for the string to lose half its energy. If the string loses energy at a constant rate, how many watts of sound are given off by the string? **Key 6**

9. Assume that no energy is given off behind the guitar and that energy is given off equally in all directions in front of the guitar. About how many watts are transferred through a square meter one meter away? **Key 33**

10. How many decibels is your answer for Helping Question 9? **Key 38**

11. If the higher-frequency string has frequency f_1 and the lower f_2, what is the frequency of their beats? **Key 40**

12. Graph the functions $\sin x$ and $\sin(2x)$ on the interval 0–π. Use these two graphs to graph their product function $\sin x \sin(2x)$ (not their dot product!) on the same interval. What is the net area under the curve? Use *symmetry*! **Keys 4 & 7**

13. The key idea is contained in Helping Question 12.

14. You know that if n is even, $A_n = 0$. If n is odd, what relation holds between

$$\frac{2}{\pi} \int_0^{\pi/2} f(x)\, dx$$

and

$$\frac{2}{\pi} \int_0^{\pi} f(x) \sin(nx)\, dx?$$

Key 21

15. You must integrate

$$2\left(\frac{2}{\pi} \int_0^{\pi/2} f(x) \sin(nx)\, dx\right),$$

which is

$$\frac{8y}{\pi^2} \int_0^{\pi/2} x \sin(nx)\, dx$$

In the usual notation of integration by parts, what should you take for u and what should you take for dv? **Key 15**

NOTES: Orders of Magnitude

The problems in the Learning Guides that you've done so far have consistently deemphasized arithmetic. As you recognize by now, knowing that the answer to a certain problem is, say, 6.5 and not 6.2, almost never adds to your understanding of the problem. However, knowing that the answer to a problem is 6×10^5 and not 6×10^2 does add to your understanding of the problem!

A rough indication of the size of a number is the power of 10 that most closely represents it. This exponent is called the number's order of magnitude. *You should know roughly the orders of magnitude of the quantities you're working with.*

This Learning Guide's notes are not suggesting that you should forget what you've learned and plug in numbers everywhere! Rather, they are advising you to try to *understand physically the magnitudes involved.* You obviously couldn't just say that the speeding physicist would be fined 120 million dollars, but you should recognize immediately that his defense is utterly ridiculous. When you check a numerical answer with the Answer Key, you should do more than just see that it matches—you should think about it for a little while and *make it part of your physical intuition.* If you were surprised that the phase velocity in a guitar string was more than half a kilometer per second, you shouldn't be surprised again! Of course, for an answer like "half a kilometer per second" to mean anything to you, you have to have some reference points. For example, you should have *no doubts* about the speed of sound being about a mile every five seconds, atmospheric pressure being about 14 lb/in^2, and water weighing about 60 lb/ft^3. To help build up your intuition about sound, you should look carefully at Figure 14-17 in Fishbane, Gasiorowicz, & Thornton.

Physicists often call a rough calculation an "order-of-magnitude" or "back-of-the-envelope" calculation. Even though they're only approximate, they can be very revealing. Problem IV about the guitar string is a good example—the crudest of calculations showed that almost all of the string's energy went into heat.

ANSWER KEY

1. $1.20f$
2. To the right
3. 220, 660, and 1100 Hz
4. The two shaded regions have equal area.

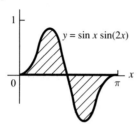

5. About 10^{-4} W/m², or 80 dB; the same as a jackhammer 10 yards away
6. About 5 s; about 5×10^{-4} W
7.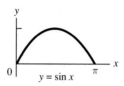
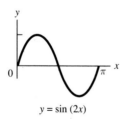

8. $2\pi^2 y_m^2 f^2 \mu$
9. $\omega = 1$ rad/s; $f = 1/2\pi$ Hz
10. 2π s
11. Every adjacent pair among the bottom 26 keys (plus some keys that are close but not adjacent in the very lowest octave)
12. $\sin(\omega t - kx)$ and $\sin(\omega t + kx)$; ω/k
13. $\omega^2 \mu \Delta \ell$
14. They're zero.
15. $u = x$; $dv = \sin(nx)\,dx$
16. The mass per unit length of the string (or the tension in the string); a wave on a thick rope (μ large) will transfer more energy than a wave with the same wave function on a thin string (μ small).
17. The intensity of the sound goes up and down, but there are no clearly defined beats since the tones aren't pure.
18. $1.25f$
19. About 0.005 J
20. 78.5 cm; 331 Hz
21. The second is twice the first.
22. 331 Hz; yes
23. 6.24 m/s (vertically)
24. $\ell = \pi/4$
25. 440 and 880 Hz
26. 1; $(0.003) \times (2080) \times [-\sin(2080\omega t)]$
27. No; yes
28. 2π m
29. $1/\Delta\ell$; y_m
30. 1 cm
31. For $n = 1, 5, 9, \ldots$; $A_n = 8y/\pi^2 n^2$.
32. $1\ \text{m}^{-1}$
33. About 8×10^{-5} W
34. ± 520 m/s, which is greater than the speed of sound in air
35. The medium (the air) is stationary in one person's frame but moving in the other's.
36. Over 120 million dollars!
37. 1 m/s
38. 79 dB
39. I is too big by about 15 or 20 dB, so about 97% to 99% of the energy must be lost to heat!
40. $f_1 - f_2$

learning guide 10

Fluid Mechanics

Suggested Reading: Fishbane, Gasiorowicz, & Thornton, Chapter 16

PROBLEM I

In the *hydraulic press* drawn in the figure, the small piston on the left has cross-sectional area a and the larger piston on the right has area A. The fluid used in the press is water.

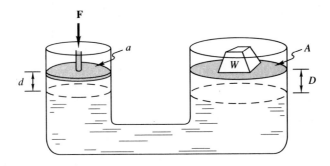

1. What force F must be applied to the small piston to hold a weight W on the large piston at the same level as the small piston? See Helping Question 1 if you need assistance. **Key 17**
2. If the force moves the left piston down a distance d, by what distance D does the weight rise? **Key 14**

3. At the levels reached after the motion of part (2), what force F' must be applied to the small piston to hold up the weight? Denote the mass density of water by ρ. Use Helping Question 2 if your answer doesn't check.
Key 18

4. In a typical hydraulic press, $a = 1$ in^2 and $A = 1$ ft^2. The *weight* of water is about 60 lb/ft^3. Give numerical answers to parts (1), (2), and (3), assuming $W = 1440$ lb and $d = 5$ ft. **Key 21**

PROBLEM II

A helium-filled balloon is anchored by a string to the floor of an enclosed railroad car. The train accelerates in the forward direction with acceleration $a = 1$ m/s^2.

1. Which way does the balloon tilt? Stumped? See Helping Question 3.
Key 6

2. What angle does the string holding the balloon make with respect to the vertical? See Helping Question 4 for a hint. **Key 3**

PROBLEM III

The figure shows a Venturi device, which is being used to create a partial vacuum. The cross section decreases from $a_1 = 20$ cm^2 at the inlet and outlet to $a_2 = 10$ cm^2 at the narrow throat in the middle. The height of the fluid in the inlet standpipe, which is open to the atmosphere at its top, is $h_1 = 0.3$ m. The standpipe at the throat is closed and is measured to be at a slight vacuum, $p_2 = 0.9$ atm. In what follows you may neglect viscous forces on the water. Take atmospheric pressure to be $p_0 = 10^5$ Pa.

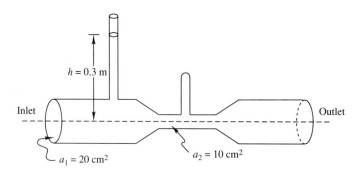

1. How much water is flowing through the pipe (i.e., what is the flux Φ through the pipe)? If you need to, consult Helping Questions 5, 6, and 7. **Key 1**

PROBLEM IV

Water flows through a hole at the bottom of a tank that is filled to a height $h = 3$ m, as shown. The radius of the hole is $r_1 = 1.5$ cm.

Learning Guide 10   Fluid Mechanics

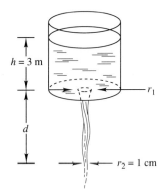

1. What is the speed of the water immediately after it leaves the hole? See Helping Question 8 if you need to. **Key 11**
2. At what distance d below the bottom of the tank is the radius of the stream reduced to $r_2 = 1$ cm? For a clue, see Helping Questions 9 and 10.
Key 25

PROBLEM V

A plastic ball has a constant density equal to one-quarter that of water. Embedded in the surface of the ball is a solid steel pellet that has mass equal to that of the rest of the ball. The ball is put into a body of water, halfway immersed, with the pellet at its highest point, as shown.

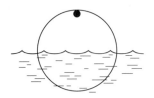

1. Make a sketch that shows the center of mass and the center of buoyancy of the ball. Show also the gravitational force and the buoyant force on the ball and their effective points of application. Reread Section 16-4 of Fishbane, Gasiorowicz, & Thornton if you're confused. **Key 28**
2. Your answer for part (1) shows that the ball is in static equilibrium. Is the equilibrium stable? **Key 7**

A solid barge is made of the same plastic and also has a small steel pellet embedded in its surface with mass equal to that of the entire rest of the barge. The barge is put into the water as shown.

3. Does the barge flop over the way the ball does? If you're just guessing, look at Helping Questions 11 and 12. **Key 2**

PROBLEM VI

A boy is tired of hauling water pail by pail from a 20-ft well. Fortunately for him, the well is on the side of a hill, so he can hook up a siphon. He buys a long garden hose with an interior diameter of $d = 0.5$ in.

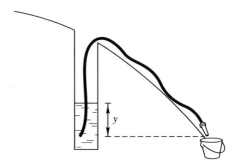

1. Having primed the siphon (filled it with water), the boy puts the free end of the hose $y = 5$ ft below the level of water in the well. How many seconds does it take him to fill his 5-gal pail? A gallon is about 200 in^3. Neglect the viscosity of the water. For a hint, see Helping Question 13. **Key 20**

   Note: By ignoring viscosity of the water, we have considerably underestimated the time.

2. Having filled his pail, the boy puts the free end 2 in above the surface level. What happens? **Key 26**

3. Just for fun, can you think of a way to leave the hose in between uses, so that (a) water doesn't flow out of the well, (b) the siphon doesn't lose its prime, and (c) conditions (a) and (b) hold even if the surface level has large fluctuations? **Key 12**

HELPING QUESTIONS

1. What physical quantity is a function only of depth in stationary liquids? **Key 4**
2. If the vertical separation between two points in a stationary liquid is h, what is the difference of the pressures at the two points? **Key 9**
3. Which way does the rest of the "heavy" air tend to move with respect to the accelerating car? **Key 10**
4. What does Einstein's equivalence principle say about the apparent direction of gravity inside the car? **Key 23**
5. What does Pascal's principle yield for the pressure at the bottom of the inlet standpipe? **Key 8**
6. What does the equation of continuity give for the ratio of the fluid speeds in the inlet (v_1) and in the throat (v_2)? **Key 15**
7. What other relation holds between the pressures and flow speeds at the inlet and throat por-

tions of the device? **Key 22**

8. What does Bernoulli's equation tell you about the relationship between the speed of the stream and the height of the water? **Key 16**

9. How are the stream radius and the flow speed related? **Key 24**

10. What is the equation for the speed of a freely falling mass? **Key 19**

11. When the ball is rotated, always remaining half immersed, does its center of buoyancy change? What about when the barge rotates? **Key 13**

12. Suppose the barge had tipped over a little bit, as shown in the diagram. Indicate the gravitational force and the buoyant force and their effective points of application. Does the net external torque tend to right the barge or flop it over? **Key 5**

13. Consider points at the surface of the well water (outside the hose) and at the opening of the hose where water is coming out. What is the pressure at these two points? Now write Bernoulli's equation for these two points, assuming that the well diameter is much larger than the hose diameter. **Key 27**

ANSWER KEY

1. $\Phi = 2.29 \times 10^{-3}$ m^3/s = 2.29 L/s
2. No
3. $\theta = \tan^{-1}(a/g) = 5.8°$
4. Pressure
5.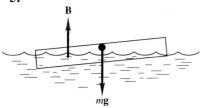

The torque tends to right the barge.

6. Forward
7. No
8. $p_1 = p_0 + \rho g h_1$
9. $\rho g h$
10. Backward
11. $v_1 = \sqrt{2gh} = 7.7$ m/s
12. Bend the hose so the last 6 or 8 ft runs back up the hill.

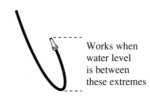

Works when water level is between these extremes

13. No; yes
14. $D = (a/A)d$
15. $v_1/v_2 = a_2/a_1$
16. $p_0 + \rho v_1^2/2 = p_0 + \rho g h$; $p_0 = 1$ atm (i.e., at both top *and* bottom)
17. $F = (a/A)W$
18. $F' = \dfrac{a}{A}W + \rho g d a \left(1 + \dfrac{a}{A}\right)$
19. $v_2^2 = v_1^2 + 2g(y_1 - y_2) = v_1^2 + 2gd$
20. 23.7 s
21. $F = 10$ lb, $D = \tfrac{5}{12}$ in, $F' = 12.1$ lb
22. $p_1 + \tfrac{1}{2}\rho v_1^2 = p_2 + \tfrac{1}{2}\rho v_2^2$

23. There is an apparent acceleration due to gravity given by $g' = \sqrt{g^2 + a^2}$ that points down and slightly backward.
24. $v_1 \pi r_1^2 = v_2 \pi r_2^2$
25. $d = h\left[(r_1/r_2)^4 - 1\right] = 12.2$ m
26. The water in the hose goes back into the well.
27. $p_1 = p_2 = 1$ atm, ignoring the small change in atmospheric pressure over 5 ft. Thus, Bernoulli's equation

$$p_1 + \rho g y = p_2 + \frac{1}{2}\rho v_2^2$$

becomes

$$\rho g y = \frac{1}{2}\rho v_2^2,$$

where v_2 is the speed of the water as it comes out of the hose.

28.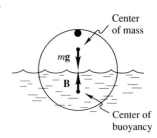

29. $F = (a/A)W$
30. Backward

learning guide 11

Thermodynamics I

Suggested Reading: Fishbane, Gasiorowicz, & Thornton, Chapters 17, 18, and 19

PROBLEM I

A 1-g block of ice is placed in a closed container holding 30 g of air. The initial temperature of the system is $-25°C$. Heat is added at the constant rate of 1 cal/s until the temperature of the system is $125°C$.

1. How much of the heat has been used to:
   a) Raise the temperature of the air *alone* from $-25°C$ to $125°C$? The specific heat of air at constant volume is 0.17 cal/g·°C. **Key 21**
   b) Raise the temperature of the ice alone from $-25°C$ to $0°C$? The specific heat of ice is very nearly 0.5 cal/g·°C. **Key 32**
   c) Melt the $0°C$ ice? You can find the numerical data for this part and parts 4 and 5 in Table 18-1 of Fishbane, Gasiorowicz, & Thornton. **Key 1**
   d) Raise the temperature of the water from $0°C$ to $100°C$? **Key 24**
   e) Boil the $100°C$ water? Assume that the air pressure is exactly 1 atm when the water reaches $100°C$ so that the water starts to boil. Assume also that all the water boils at exactly $100°C$ (actually, some won't boil until a higher temperature because of the steam pressure). **Key 17**
   f) Raise the temperature of the steam from $100°C$ to $125°C$? The specific heat of steam at pressures near atmospheric pressure is very nearly 0.5 cal/g·°C. **Key 20**

2. Make a graph of the temperature of the system as a function of time. Express time in minutes in order to avoid large numbers. **Key 33**

3. If the energy used in parts (1c) through (1e) (turning 0°C ice into 100°C steam) were used instead to lift the gram of ice, how many meters above the earth's surface would the ice be lifted? **Key 10**

PROBLEM II

A 30-g silver teaspoon at room temperature (20°C) is placed into 200 g of hot tea (100°C). No heat escapes the spoon-tea system. You may assume that the specific heat of tea is the same as that of water. The specific heat of silver is 0.0564 cal/g·K.

1. What is the equilibrium temperature of the system? Helping Questions 1 and 2 will get you going. **Key 2**

PROBLEM III

On a nice summer day the air temperature at the surface of the earth is 300 K.

1. What is the average (rms) speed of the nitrogen molecules? Of the oxygen molecules? Use Helping Question 3 if you're stumped. **Key 5**

PROBLEM IV

The mean free path of air molecules at 0°C and 1 atm is about 2×10^{-5} cm.

1. What is the effective molecular diameter of an air molecule? **Key 8**
2. At what pressure (at 0°C) would the mean free path be equal to the effective molecular diameter? **Key 3**

PROBLEM V

Work out a semiquantitative explanation of why the earth has an atmosphere but the moon does not. Some useful data: Radius of the earth $R_e = 6.4 \times 10^6$ m, radius of the moon $R_m = 1.7 \times 10^6$ m, local accelerations due to gravity $g_m \simeq g/6$ (remember Neil Armstrong), and daytime temperature of the moon $T_m = 200°F$. Clueless? See Helping Questions 4 and 5. **Key 9**

PROBLEM VI

One mole of a nonideal gas is put through the four-stage process as indicated in the accompanying p–V diagram: It begins at roughly atmospheric conditions

(a); heat is added at constant volume, causing an increase in pressure ($a \to b$); more heat is added and the gas expands at just the right rate to keep the pressure constant ($b \to c$); and heat is dumped and work is done on the gas as indicated in the diagram in stages $c \to d$ and $d \to a$ to complete the cycle.

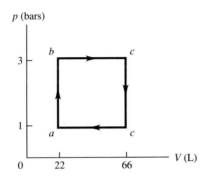

1. How much work is done during the cycle? Look at Helping Questions 6 and 7 if you can't figure out the answer. **Key 27**
2. What is the ratio

$$\frac{\text{(Work done)}}{\text{(Heat taken in)} - \text{(heat dumped)}}$$

for this cycle? Helping Question 8 will remind you of the key idea.
Key 12

PROBLEM VII

Heat is applied to an ideal diatomic gas, which is maintained at constant pressure by a piston loaded with sand. As heat is added the gas expands from an initial volume $V_i = 1\,\text{L} = 10^{-3}\,\text{m}^3$ to a final volume $V_f = 3\,\text{L}$. The initial pressure and temperature of the gas are $p_i = 3$ atm and $T_i = 0°\text{C}$.

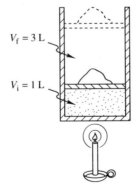

1. How many moles of gas are in the container? Use Helping Question 9.
Key 26
2. How much work is done by the gas on the piston during the expansion?
Key 11

3. What is the final temperature of the gas? **Key 16**
4. What heat Q was required? Use Helping Question 10. **Key 18**
5. How much heat went into an increase in the total internal energy of the diatomic gas? **Key 13**
6. How much of the increase in internal energy was used to increase the translational motion of the diatomic gas? **Key 19**
7. Where did the remainder of the increase in internal energy go? **Key 4**
8. Where did the remainder of the heat go? **Key 28**

HELPING QUESTIONS

1. What is the relation between the heat lost by the spoon and the heat absorbed by the tea? **Key 14**
2. Can you write an equation for the answer to Helping Question 1 in terms of the initial temperatures T_{spoon} and T_{tea}, the final temperature T, and other constants? **Key 6**
3. What is the relation between average translational kinetic energy and temperature? **Key 31**
4. What is the escape velocity for a particle on the earth's surface? **Key 25**
5. Whose distribution gives the probability that a molecule will attain a given speed? **Key 30**
6. In geometrical terms, what is the work done by the process drawn? **Key 15**

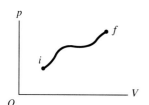

7. In the same geometrical terms as Helping Question 6, what is the work done by the four-stage process in the problem? **Key 7**
8. What does the first law of thermodynamics say? What is the change in internal energy of the gas after one cycle? **Key 23**
9. What does the ideal gas law say? What is atmospheric pressure in N/m^2? **Key 29**
10. What is the molar heat capacity of a diatomic gas at constant pressure? **Key 22**

Notes—Temperature, Internal Energy, and the Connection between Them

Most students have more trouble understanding thermodynamics than they have understanding mechanics. If you belong to this group, think about why you're having trouble with thermodynamics. It can't be that you're having more trouble with the *math*. In fact, you need to know less math for thermodynamics than you did for mechanics—there are no vectors in thermodynamics, not

Learning Guide 11   Thermodynamics I

even any trigonometry. It must be that you're having more trouble with the *physics*. Everything will fall into place once you understand the three key physical concepts in the suggested reading for this Learning Guide—*temperature, internal energy, and the connection between them*. These three concepts are easiest to understand from the point of view of kinetic theory.

Problem I showed you the necessity of making a distinction between temperature and internal energy. Even though the internal energy of the water-air system was increased at a constant rate, the temperature did not simply increase proportionally. Over certain time intervals the temperature change was proportional to the internal energy change. But the constant of proportionality was different for different intervals; overall, the dependence was rather complicated.

In Problem II you had to use all three concepts to find the equilibrium temperature of the spoon-tea system. Let's look at a formal solution of this problem. Call the final temperatures $T_{f,\text{spoon}}$ and $T_{f,\text{tea}}$. Then if we introduce the extra unknowns ΔU_{spoon} and ΔU_{tea}, we have four unknowns, so we need four equations:

$$T_{f,\text{spoon}} = T_{f,\text{tea}}$$

(which comes from the zeroth law, whose basic concept is temperature)

$$\Delta U_{\text{spoon}} = -\Delta U_{\text{tea}}$$

(which comes from the first law, whose basic concept is internal energy)

$$\Delta U_{\text{spoon}} = m_{\text{spoon}} c_{\text{spoon}} (T_{f,\text{spoon}} - T_{\text{spoon}}) \quad \text{and} \quad \Delta U_{\text{tea}} = m_{\text{tea}} c_{\text{tea}} (T_{f,\text{tea}} - T_{\text{tea}})$$

(which come from the definition of specific heat, whose basic concept is the connection between T and U)

Let's turn now to kinetic theory and see what light it sheds on the three key concepts. You learned that the temperature of a substance macroscopically at rest is proportional to the average translational kinetic energy of the molecules (this is exactly true for ideal gases and approximately true in other cases). You have to distinguish between temperature and internal energy because of the qualifier "translational kinetic." Why is this type of energy important? When molecules collide they tend to equalize their translational kinetic energies. So if two different types of molecules are in a container, their random collisions serve to even out the average translational kinetic energy but not the average total energy; this is just the content of the zeroth law!

As you know, kinetic theory takes all the mystery out of the first law. The U introduced in Chapter 18 is just the internal energy of the system. So, for example, you were just using conservation of energy in Problem VII, part (2).

Finally, the connection between temperature and internal energy—heat capacities—is made much less mysterious by kinetic theory. The connection is made by the equipartition of energy. You can get very good estimates for molar heat capacities at constant volume just by counting degrees of freedom. As the text discusses, the molar heat capacities at constant volume for monatomic and diatomic gases are $\frac{3}{2}R$ and $\frac{5}{2}R$, respectively. When the situation is more complicated, you can still use the equipartition theorem qualitatively. Problem I said ice and steam have lower molar heat capacities than water. Can you show that this is reasonable by using the equipartition theorem? Which terms go away? You can get molar heat capacities at constant pressure easily too: for gases, $C_P \simeq C_V + R$, and for solids and liquids, $C_P \simeq C_V$. So if you memorize the values

$$R \simeq 8 \frac{\text{J}}{\text{mol} \cdot \text{K}} \simeq 2 \frac{\text{cal}}{\text{mol} \cdot \text{K}},$$

you can do many numerical thermodynamics problems quite accurately by using the equipartition theorem instead of tables.

ANSWER KEY

1. 80 cal
2. Between 99 and 100°C
3. 1000 atm
4. Into rotational motion
5. 517 m/s; 483 m/s
6. $m_{\text{tea}} c_{\text{tea}} (T_{\text{tea}} - T) = m_{\text{spoon}} c_{\text{spoon}} (T - T_{\text{spoon}})$
7. The area of the square
8. 2×10^{-8} cm
9. Compare the quantity

    $P' = \exp\left(-M_{N_2} v_{\text{esc}}^2 / RT\right)$

    for the earth,

    $P' = \exp\left(-g R_e M_{N_2} / RT\right)$
    $\simeq e^{-700} \simeq 10^{-300}$,

    versus the moon,

    $P' = \exp\left(-g R_M M_{N_2} / 6RT\right)$
    $\simeq e^{-25} \simeq 10^{-11}$.

    Note that 10^{289} is a big number!
10. 3.07×10^5 m
11. 606 J
12. 1
13. 1509 J
14. They are equal.
15. The area under the curve
16. 819 K
17. 539 cal
18. 2113 J
19. 905 J
20. 12.5 cal
21. 765 cal
22. $7R/2$
23. $\Delta U = Q - W$; 0
24. 100 cal
25. $v_{\text{esc}} = \sqrt{2 g R_e}$
26. 0.133 mol
27. 8800 J
28. Into the work done by the gas
29. $PV = nRT$; 1.013×10^5 N/m^2
30. Maxwell's speed distribution (see Fishbane, Gasiorowicz, & Thornton, Section 19-5)
31. $E_{\text{av}} = \frac{1}{2} M v_{\text{rms}}^2 = \frac{3}{2} RT$
32. 12.5 cal
33.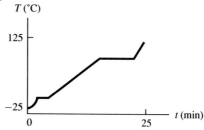

learning guide 12

Thermodynamics II

Suggested Reading: Fishbane, Gasiorowicz, & Thornton, Chapters 20 and 21

PROBLEM I

The *p–V* diagram shown is the same as the one used in Learning Guide 11, Problem VI. There you calculated the ratio

$$\frac{\text{(Work done)}}{\text{(Heat taken in)} - \text{(heat dumped)}}$$

in a cycle to be equal to unity, independent of the type of gas used in the engine. As you know now, a more important ratio is the efficiency of the engine,

$$\eta = \frac{\text{(work done)}}{\text{(heat taken in)}},$$

since the amount of fuel used to produce the work depends on (heat taken in) and not on (heat taken in) − (heat dumped).

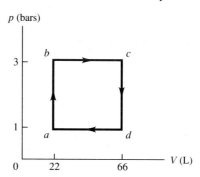

1. What is the efficiency of the engine if the working substance of the engine is one mole of ideal monatomic gas? One mole of ideal diatomic gas? If your answer doesn't check after a good try, look at Helping Questions 1 through 4. **Key 8**

PROBLEM II

A gas is put through a Carnot cycle whose graph on a p–V diagram contains the points labeled a and c in Problem I. The gas starts at point a. It is compressed adiabatically until its temperature is the same as the temperature of the gas when it is at point c. It is then expanded isothermally until it reaches point c. It is then expanded adiabatically until its temperature is the same as it was at point a. Finally, it is compressed isothermally until it returns to point a.

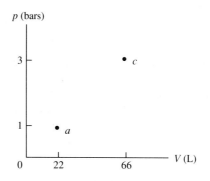

Assume that the gas is ideal and monatomic.

1. What is the efficiency of the engine? Stuck? Use Helping Question 5. **Key 9**

2. Indicate the two other corners of the cycle on a p–V diagram. This is somewhat involved. Use Helping Question 6. **Key 21**

Perhaps now you have a better idea of the difficulty of designing a heat engine with near 100% efficiency! Here are two more quick questions to solidify your understanding of heat engines.

3. If the gas used in the Carnot cycle were ideal and diatomic, would the efficiency of the engine be the same as in part (1)? Would the plot of its cycle on a p–V diagram be the same as in part (2)? **Key 26**

4. The cycle of Problem I can be approximated arbitrarily well by Carnot cycles. Why is its efficiency not the same as the efficiencies calculated in parts (1) and (3)? **Key 33**

PROBLEM III

Three thousand calories of heat are added to an ice cube at 0°C, turning it into water, also at 0°C.

1. What is the change in entropy, $S_f - S_i$? If your answer doesn't check, reread Section 20-5 of Fishbane, Gasiorowicz, & Thornton. **Key 1**

   A rock of mass m has a constant specific heat c (that is, the specific heat is independent of temperature). The rock is heated from T_i to T_f.

2. What is the change in entropy $S_f - S_i$? Use Helping Questions 7 and 8 as needed. **Key 30**

   An ideal gas of n mol undergoes an isothermal expansion from a volume V_i to a volume V_f at temperature T.

3. What is the change in entropy, $S_f - S_i$? If you're stuck after a good effort, use as few of Helping Questions 9 through 11 as possible. **Key 15**

PROBLEM IV

One mole of monatomic ideal gas is taken from an initial state i to a final state f by two different processes as indicated in the p–V diagram. Process 1 follows the solid path, whereas process 2 follows the dotted path. The curved line is an isotherm. In terms of p, V, and the initial temperature T, calculate for each process:

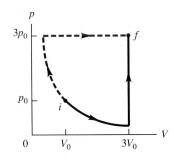

1. The increase in internal energy. See Helping Question 12. **Key 10**
2. The work done by the gas. See Helping Question 13. **Key 22**
3. The heat added to the gas. See Helping Question 14. **Key 16**
4. The increase in entropy. If you're stuck, look at Helping Question 15. **Key 2**

PROBLEM V

The rock and ideal gas of Problem III are each involved in an *irreversible* process. The rock, of mass m_1 and constant specific heat c_1, starts at a temperature T_1. It is placed in contact with a cooler rock of mass m_2, constant specific heat c_2, and temperature T_2. There is a rapid transfer of heat across the boundary, and the rocks reach their equilibrium temperature. No heat is transferred from the rocks to the surroundings, and the rocks do no work.

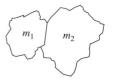

1. What is the increase in the entropy of the system? Use Helping Questions 16 and 17. **Key 42**

One mole of an ideal gas is in the left chamber of the container sketched here. The valve at the narrow passage is opened, and the gas flows rapidly and turbulently to its equilibrium point. At equilibrium, of course, it is spread evenly throughout the two chambers. Again, no heat is transferred from the gas to its surroundings, and no work is done by the system.

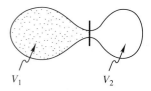

2. What is the increase in entropy of the system? If you're lost, use Helping Questions 18 and 19. **Key 39**

PROBLEM VI

1. Without looking at the text, *derive* the efficiency of a Carnot cycle working between T_H and T_C. Start from the equation

$$\text{(Entropy taken in)} = \text{(entropy dumped)},$$

and express your answer in terms of T_H and T_C. See Helping Question 20. **Key 3**

PROBLEM VII

A steel rod of length 2 m at 0°C is placed in boiling water and allowed to reach equilibrium.

Learning Guide 12   Thermodynamics II

1. What is its new length? If your answer doesn't check, use Helping Question 21. **Key 23**

Suppose that the two ends of the rod were constrained to stay exactly 2 m apart. Because of the thermal expansion, the bar bends at its center, as shown in the diagram.

2. By what distance d does the center of the bar rise? Stuck? Look at Helping Question 22. **Key 38**

PROBLEM VIII

A small ball of radius R and temperature $5°C$ is observed to be at rest just below the surface of a large pool of water that is maintained at $55°C$. Now the ball warms up and comes to thermal equilibrium with the water. What fraction of the volume of the ball is out of the water? (The coefficient of linear expansion of the material of the ball is $\alpha_{ball} = 10^{-4}/°C$.) Use Helping Question 23. **Key 7**

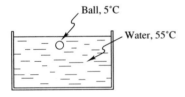

PROBLEM IX

A double-pane storm window of area A consists of two panes of glass separated by an air space. The thermal conductivity of the air is k_a, that of the glass is k_g. The pane thickness is l_g and the air thickness is l_a.

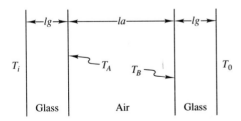

1. If the outside temperature is T_o, which is less than the inside temperature T_i, what is the heat loss per second through the window? **Key 43**
2. What are the temperatures at the inner surfaces of the panes of glass, T_A and T_B? If you need help, see Helping Questions 24 and 25. **Key 13**

PROBLEM X

If you look at Table 18-2 of Fishbane, Gasiorowicz, & Thornton, you will see that the molar heat capacities of metals are *roughly* constant, independent of the type of metal. This is a consequence of the equipartition of energy, discussed in Section 19-6 in the context of gases. Just as hard spheres in random motion form a reasonably good model for monatomic gases, hard spheres in a lattice connected by massless springs form a reasonably good (but not perfect) model for metals. There are 6.02×10^{23} atoms in a mole of metal, so only a small part of the lattice is drawn to the right.

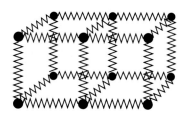

1. If there are N atoms, how many terms are there in the expression for the total energy? Because the lattice is so big, you can neglect edge effects. Use Helping Questions 26 and 27. **Key 20**
2. Use what you know about equipartition of energy to find the molar heat capacity at constant volume of a metal. **Key 46**

The calculation above yields molar heat capacities at *constant volume*. What about the heat capacities for metals at *constant pressure*? It turns out that this distinction isn't important; for liquids and solids, $C_p \simeq C_V$, because liquids and solids do negligible work when they expand.

3. What is the *molar* heat capacity of water in cal/mol·K? For comparison, also give your answer in terms of R. What is the molar heat capacity for metals and gases? (Use equipartition of energy qualitatively.) **Key 31**

HELPING QUESTIONS

1. What are the temperatures T_a, T_b, T_c, and T_d at the corners of the diagram? Along which legs of the cycle is heat taken in? **Key 37**
2. What is the molar heat capacity at constant volume for a monatomic ideal gas? For a diatomic ideal gas? Then how much heat is added in $a \to b$ in each of the two cases? **Key 12**
3. What is the molar heat capacity at constant pressure for a monatomic ideal gas? For a diatomic ideal gas? Then how much heat is added in $b \to c$ in each of the two cases? **Key 28**

Learning Guide 12 Thermodynamics II

4. What is the work done in the cycle? **Key 35**
5. What is the efficiency of a Carnot cycle working between temperatures T_H and T_C? **Key 5**
6. The cycle is plotted *very inaccurately*. An exact plot can be made because you can find *equations* describing all four curves. The equation for $b \to c$ is $pV = p_b V_b$. You can get this from the ideal gas law, since $b \to c$ is an *isotherm*. Find equations for the other three curves. Then use these equations to find the two other corners. **Key 47**

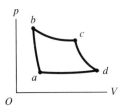

7. What is dS in terms of dQ and T? What is the definition of the specific heat c? Combine these two equations into one that doesn't involve dQ. **Key 34**
8. Express the change of entropy $S_f - S_i$ as an integral. Can you evaluate this integral by using the equation from Helping Question 7? **Key 18**
9. Express the first law $dU = dQ - dW$ in terms of U, T, S, p, and V. Now solve your expression for dS. **Key 14**
10. Since the gas is ideal and the process is isothermal, what is dU? **Key 6**
11. Express the change in entropy $S_f - S_i$ as an integral. Can you evaluate this integral by using the answers to Helping Questions 9 and 10 and the ideal gas law, $pV = nRT$? **Key 32**
12. What is the formula for the internal energy of a monatomic ideal gas? **Key 44**
13. $dW = ?$ **Key 19**
14. What is the first law of thermodynamics? **Key 29**
15. Use Helping Questions 9 and 12 to find an equation for the change of entropy of an ideal monatomic gas. **Key 36**
16. What is the relation between the heat released by rock 1 and the heat absorbed by rock 2? Can you deduce the equilibrium temperature from this? **Key 4**
17. Use your result from Problem III, part (2).
18. Can you think of a *reversible* process that has the same initial and final states as the irreversible process described? **Key 25**
19. Use your result from Problem III, part (3).
20. Can you express the equation (entropy taken in) = (entropy dumped) in terms of T_H, T_C, Q_H, and Q_C? Now use the definition of efficiency and the first law of thermodynamics. **Key 45**
21. What is the definition of the coefficient of linear expansion α? Use Table 21-3 of the text to find the value of α for steel. **Key 17**
22. To what right triangle can you apply the Pythagorean theorem? **Key 40**
23. What is the relation between the coefficient of volume expansion β and α? **Key 27**
24. When the heat flow has reached a steady state, what is the relation between the heat flow H_i reaching surface A from the inside and the heat flow H_o leaving surface A toward the outside? **Key 24**
25. What are H_i and H_o? **Key 11**

26. How many terms are there in the expression for the kinetic energy? **Key 48**

27. About how many springs are there? Then how many terms are there in the expression for the potential energy? **Key 41**

Notes—Entropy

In the notes from Learning Guide 11 you saw how the microscopic viewpoint can give you a deeper understanding of the thermodynamic variables T and U and their associated laws—the zeroth and first laws. In this learning guide you'll see how the microscopic viewpoint can give you a deeper understanding of the thermodynamic variable S and its associated law—the second law.

The key to this understanding is interpreting the entropy of a system as the disorder of the system. Let's look at two examples from Problem III. When ice melts, the H_2O molecules go from being regularly arranged in a crystal lattice to being in random motion. Intuitively, the disorder of the system increases, and, sure enough, the entropy change you calculated was positive. When a gas expands, the molecules go from being localized in a relatively small volume to being spread out over a larger volume. Again, intuition says the disorder increases, and again the entropy change you calculated was positive. With this interpretation, the second law says that the disorder of the universe never decreases. So it's not very hard to turn an abstract concept into an intuitive one—just replace the word "entropy" with the word "disorder"!

Entropy is intimately related to heat through the equation $dS = dQ/T$. When you add heat to a system you are increasing its entropy. This makes sense from the point of view of disorder: when you add heat, you are adding energy in a disordered form. But you must be careful to distinguish heat from entropy. As you saw in Problem IV, it doesn't make sense to talk about "heat energy" in a system. It *does* make sense to talk about the entropy or disorder in a system. This is the importance of the innocent-looking T in the equation $dS = dQ/T$. If you add a certain amount of heat dQ to a system, the amount of entropy you add depends on the temperature. Because of this factor T, S is a state variable.

If you take a course in statistical mechanics, you will see that thinking in terms of disorder doesn't just happen to give you the right answers; thinking in terms of disorder works because "disorder" really does describe what's physically going on. The basic idea is this: when a system is in equilibrium in a fixed macroscopic state (for a gas, p, V, and T would be fixed), it is microscopically changing state constantly. *Some macroscopic states have more microscopic states corresponding to them than others have.* As an example, consider the gas of Problem V. There are more microscopic states corresponding to the final state (large volume) than there are corresponding to the initial state (small volume). In quantum statistical mechanics, there is actually a number N of possible microscopic states corresponding to a given macroscopic state. The entropy of the macroscopic state is just $k \ln N$! Here k is Boltzmann's constant, which is present only for dimensional reasons. So, quite literally, entropy measures disorder!

ANSWER KEY

1. 11 cal/K
2. Process 1:

   $\Delta S = R \ln 3 + \frac{3}{2} R \ln 9 = 4R \ln 3$

   Process 2:

   $\Delta S = -R \ln 3 + \frac{5}{2} R \ln 9 = 4R \ln 3$

3. $\eta_C = 1 - \dfrac{T_C}{T_H}$
4. They're equal;

   $T = \dfrac{c_1 m_1 T_1 + c_2 m_2 T_2}{c_1 m_1 + c_2 m_2}.$

5. $\eta_C = 1 - \dfrac{T_C}{T_H}$
6. $dU = (\text{constant})\, dT = 0$
7. 1.5%
8. Monatomic: 22%
   Diatomic: 15%
9. 89%
10. Process 1: $12RT$
    Process 2: $12RT$
11. $H_i = \dfrac{A k_g (T_i - T_A)}{l_g}$

    $H_o = \dfrac{A k_a k_g (T_A - T_o)}{k_g l_a + k_a l_g}$

12. $\frac{3}{2}R$; $\frac{5}{2}R$

    $Q_{\text{mono}} = \dfrac{3}{2}(p_b V_b - p_a V_a);$

    $Q_{\text{dia}} = \dfrac{5}{2}(p_b V_b - p_a V_a).$

13. $T_A = \dfrac{(k_a l_g + k_g l_a) T_i + k_a l_g T_o}{k_g l_a + 2 k_a l_g}$

    $T_B = \dfrac{k_a l_g T_i + (k_a l_g + k_g l_a) T_o}{k_g l_a + 2 k_a l_g}$

14. $dU = T\, dS - p\, dV$

    $dS = \dfrac{dU}{T} + \dfrac{p\, dV}{T}$

15. $\Delta S = nR \ln \dfrac{V_f}{V_i}$

16. Process 1: $pV(12 + \ln 3)$
    Process 2: $pV(20 - \ln 3)$
17. $\alpha = \dfrac{1}{\Delta T}\left(\dfrac{\Delta l}{l}\right)$
18. $\displaystyle\int_{S_i}^{S_f} dS$; yes: $\displaystyle mc \int_{T_i}^{T_f} \dfrac{dT}{T}$
19. $p\, dV$
20. $6N$
21.

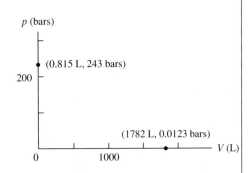

22. Process 1: $pV \ln 3$
    Process 2: $pV(8 - \ln 3)$
23. 2.0022 m
24. $H_i = H_o$
25. The gas expands quasi-statically, doing work against a very slowly moving wall. Enough heat is added so that the internal energy remains constant.
26. Yes; no
27. $\beta \simeq 3\alpha$
28. $\frac{5}{2}R$; $\frac{7}{2}R$

    $Q_{\text{mono}} = \frac{5}{2}(p_c V_c - p_b V_b)$

    $Q_{\text{dia}} = \frac{7}{2}(p_c V_c - p_b V_b)$

29. $dU = dQ - dW$
30. $\Delta S = mc \ln \dfrac{T_f}{T_i}$
31. 18 cal/mol·K or about $9R$; besides the translational and rotational kinetic energies of the individual molecules, there is also energy binding the molecules to each other.

32. $\int_{S_i}^{S_f} dS = nR \int_{V_i}^{V_f} \frac{dV}{V}$

33. Most of the Carnot cycles are working between temperatures less extreme than T_a and T_c.

34. $dS = \frac{dQ}{T}$

and

$c = \frac{1}{m} \frac{dQ}{dT}$,

so

$dS = mc \frac{dT}{T}$.

35. $(p_b - p_a)(V_c - V_a)$

36. $S_f - S_i = nR \ln \frac{V_f}{V_i} + \frac{3}{2} nR \ln \frac{T_f}{T_i}$

37. $T = pV/R$; $T_a = 265$ K; $T_b = 794$ K; $T_c = 2380$ K; $T_d = T_b$. Heat is taken in along the legs $a \to b$ and $b \to c$.

38. 4.7 cm

39. $R \ln \left(\frac{V_1 + V_2}{V_1} \right)$

40. The triangle formed by one half of the sagging wire, one half of the horizontal distance between the ends of the wire, and the vertical droop.

41. $3N$; $3N$

42. Entropy increase =

$m_1 c_1 \ln \frac{T}{T_1} + m_2 c_2 \ln \frac{T}{T_2}$

where

$T = \frac{m_1 c_1 T_1 + m_2 c_2 T_2}{m_1 c_1 + m_2 c_2}$.

43. $H = \frac{A(T_i - T_o)}{(2 l_c/k_g) + (l_a/k_a)}$

44. $U = \frac{3}{2} nRT$

45. $\frac{Q_C}{T_C} = \frac{Q_H}{T_H}$

(entropy is constant along the adiabatic legs).

46. $3R$

47. $a \to b$: $pV^\gamma = p_a V_a^\gamma$,
  $\gamma = C_p/C_V$
  $c \to d$: $pV^\gamma = p_c V_c^\gamma$
  $d \to a$: $pV = p_d V_d$

48. $3N$

49. About $1.1 \times 10^{-6}\,°C$

50. $3N$; $3N$

51. $dU = T dS - p\, dV$

$dS = \frac{dU}{T} + \frac{p\, dV}{T}$

learning guide 13

Electricity

Suggested Reading: Fishbane, Gasiorowicz, & Thornton, Chapters 22, 23, and 24

Note: There is nothing unfamiliar about Coulomb's law if you replace

$$\mathbf{F}_{\text{grav}} = -G \frac{m_1 m_2}{r^2} \hat{\mathbf{r}}$$

with

$$\mathbf{F}_{\text{elec}} = \frac{1}{4\pi \epsilon_0} \frac{q_1 q_2}{r^2} \hat{\mathbf{r}}$$

where

$$\hat{\mathbf{r}} = \frac{\mathbf{r}}{|r|}.$$

PROBLEM I

1. One coulomb is a very large charge. What fraction of electrons must be removed from a penny in order to have a charge of $+1$ C? See Helping Questions 1 and 2. **Key 1**

2. What force would be exerted by two such pennies separated by 0.1 m? (Express your answer not only in newtons, but in terms of some "imaginative" units such as the weight of a number of pyramids of Cheops, or the force on the moon due to the earth, for example.) **Key 6**

PROBLEM II

1. Charges are located on the corners of a square as shown. If $q = 10^{-7}$ C and $a = 5.0$ cm, what is the resultant force (magnitude and direction) on the charge at the lower left? See Helping Question 3 for help with the procedure. **Key 10**

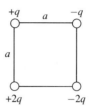

2. What is the electric field (magnitude and direction) at that point? **Key 18**

PROBLEM III

1. One of the greatest obstacles to the millennium of energy production by nuclear fusion is Coulomb's law. In order to fuse two deuterons (^{2}H) of radius $\simeq 2 \times 10^{-13}$ cm into one helium atom (^{4}He), one must overcome this force. How much work must be done to bring a pair of deuterons from infinite separation to a separation of 2×10^{-13} cm? **Key 3**
2. At approximately what temperature do the deuterium nuclei have just enough kinetic energy to overcome this "potential barrier"? Helping Questions 4 and 5 may be needed. **Key 19**

PROBLEM IV

Point charge A with a charge of 2×10^{-7} C and point charge B with a charge of 5×10^{-8} C are 15 cm apart.

1. What is the electric field at A due to charge B? **Key 23**

If your answer is incorrect, review Fishbane, Gasiorowicz, & Thornton, Section 22-3.

2. What is the electric field at B due to charge A? **Key 2**
3. What is the magnitude of the force on charge A? **Key 7**
4. What is the magnitude of the force on charge B? **Key 15**
5. Look at your answers above. Why are the forces equal? **Key 26**

Learning Guide 13   Electricity

PROBLEM V

1. A total charge Q is spread uniformly along a rod of length l. Find the electric field **E** at the point P a distance y away from one end of the rod. If you need help, see Helping Question 6. **Key 21**

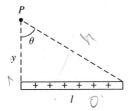

2. What does your answer become as $l \to \infty$ and Q/l remains constant? **Key 13**

3. Show that as $l \to 0$, your answer converges to that for a point charge.

PROBLEM VI: MILLIKAN'S OIL DROP EXPERIMENT

R. A. Millikan built an apparatus in which a tiny charged oil drop, placed in an electric field **E**, could be "balanced" by adjusting **E** until the electric force on the drop was equal and opposite to its weight. If the radius of the drop is 1.59×10^{-4} cm, **E** at balance is 1.92×10^5 N/C, and the density of the oil is $\rho = 0.851$ g/cm^3.

1. What charge is on the drop? **Key 16**
2. Why did Millikan not try to balance electrons in his apparatus instead of oil drops? (Millikan first measured the electronic charge in this way. He measured the drop radius by observing the limiting speed that the drops attained when they fell in air with the electric field turned off. He charged the oil drops by irradiating them with bursts of X-rays.) **Key 5**
3. In an early run (1911) Millikan observed that the following measured charges, among others, appeared at different times on a single drop:

   6.563×10^{-19} C 13.13×10^{-19} C 19.71×10^{-19} C
   8.204×10^{-19} C 16.48×10^{-19} C 22.89×10^{-19} C
   11.50×10^{-19} C 18.08×10^{-19} C 26.13×10^{-19} C

   What value for the elementary charge e can be deduced from these data? **Key 22**

PROBLEM VII

The following questions test your understanding of Gauss' law. If you have difficulty with any of them, study Fishbane, Gasiorowicz, & Thornton, Chapter 24, again.

1. A point charge is placed at the center of a spherical Gaussian surface. Is the electric flux Φ_E changed:
   a) If the surface is replaced by a cube of the same volume? **Key 50**
   b) If the sphere is replaced by a cube of one-third the volume? **Key 25**
   c) If the charge is moved off center in the original sphere, still remaining inside? **Key 24**
   d) If the charge is moved just outside the original sphere? **Key 31**
   e) If a second charge is placed near, and outside, the original sphere? **Key 38**
   f) If a second charge is placed inside the Gaussian surface? **Key 52**
2) A surface encloses an electric dipole of moment p. What can you say about Φ_E for this surface? **Key 30**
3) Suppose that a Gaussian surface encloses zero *net* charge. Does Gauss' law require that **E** equal zero for all points on the surface? Is the converse of this statement true? That is, if **E** equals zero everywhere on the surface, does Gauss' law require that there be no net charge inside? **Key 20**
4) A plane surface of area A is inclined so that its normal makes an angle α with a uniform field **E**. Calculate Φ_E for this surface. **Key 32**

PROBLEM VIII

The electric field components in the figure are $E_x = a\sqrt{x}$ and $E_y = E_z = 0$.

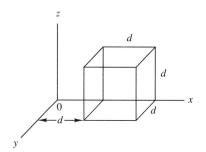

Calculate:

1. The electric flux Φ_E through the cube. **Key 4**
2. The charge within the cube. **Key 35**

If you have trouble with part (1), see Helping Questions 7 and 8; with part (2), see Helping Question 9.

PROBLEM IX

A thin metallic spherical shell of radius R_1 carries a charge q_1. Concentric with it is another thin metallic spherical shell of radius R_2 ($R_2 > R_1$) carrying charge q_2. Use Gauss' law to find the electric field at radial points r where

Learning Guide 13 Electricity

1. $r < R_1$. **Key 53**
2. $R_1 < r < R_2$. **Key 45**
3. $r > R_2$. **Key 36**
4. How is the charge on each shell distributed between the inner and outer surfaces of that shell? **Key 39**

If you have trouble with parts (1), (2), or (3), use as many of Helping Questions 10 through 13 as necessary. For part (4) use Helping Questions 14 to 16.

PROBLEM X

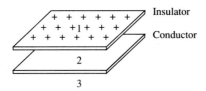

A large plate of insulating material of area A carries a total charge Q_1, which is uniformly distributed over the area of the plate. Beside this plate and parallel to it is a plate of the same size and shape of a conducting material.

Find the electric fields for regions 1, 2, and 3 for the following circumstances:

1. Conductor isolated and having zero net charge; **Key 12**
2. Conductor isolated and carrying total charge Q_2 **Key 43**

Express your answers in terms of Q_1, Q_2, A, and ϵ_0.
If you have difficulties, try Helping Questions 17 through 20.

PROBLEM XI

Suppose that an electric field exists in a region of space where there are no charges; that is, suppose that the electric field is produced by charges that are outside the region considered. Furthermore, suppose that at a certain point P in this region, the electric field is zero. Therefore, if a test charge $q_0 > 0$ is placed at P, there will be no force acting on it. If q_0 is moved slightly away from P in a certain direction, it is found that the electric field exerts on q_0 a force directed back toward P. Show that there *must* also exist a direction such that if q_0 is moved slightly away from P in this new direction, then the electric field will exert on q_0 a force directed *away from* P. (Try to use Gauss' law.) Thus, there can be no stable equilibrium in a static electric field.

If you've constructed a proof, then it must be right. You are finished. If not, look at Helping Questions 21 and 22.

HELPING QUESTIONS

1. How many electrons are in one coulomb? **Key 17**
2. How many electrons are in a penny? (It is useful to remember that a penny weighs about 3 g, and that $A_{Cu} = 64$ and $Z_{Cu} = 29$.) **Key 8**
3. Find the y-component of force on $+2q$ due to $+q$ above it (and continue with the other two charges, and the x-components). **Key 9**
4. Using an analogue to gravitational potential energy, what is the magnitude of the energy barrier represented by this force? **Key 11**
5. What do the conservation of energy and the equipartition of energy say to help you here? **Key 14**
6. Did you read Fishbane, Gasiorowicz, & Thornton, Section 23-3?
7. What is the flux through the walls of the cube that are perpendicular to the z-axis? Through the walls of the cube that are perpendicular to the y-axis? **Key 28**
8. What is the flux through the wall of the cube that is perpendicular to the x-axis at $x = d$? Through the wall of the cube that is perpendicular to the x-axis at $x = 2d$? **Key 33**
9. How is the total flux through the cube related to the charge inside and outside the cube? **Key 49**
10. What is a convenient Gaussian surface to which to apply Gauss' law? **Key 27**
11. What is the total charge enclosed by your Gaussian surface? **Key 44**
12. What does the symmetry of the problem tell you about the electric field? **Key 51**
13. Express the flux through your Gaussian surface in terms of the electric field. **Key 41**
14. For each shell, consider a concentric spherical Gaussian surface which lies between the inner and outer surface of the given conducting shell. What is the flux through this Gaussian surface? **Key 29**
15. For the results of Helping Question 14, calculate the *total* charge enclosed by the Gaussian surface. **Key 34**
16. What does charge conservation tell you about the *total* charge on each shell? **Key 37**
17. With the insulator *alone* present, what is the field "above" and "below" the plate? **Key 46**
18. The "downward" field is intersected by the conductor. What induced charge density (i) terminates the field? (ii) Regenerates the field at the bottom face of the conductor? **Key 42**
19. Now take the insulator, reisolate the conductor, and deposit charge Q_2 on it. With the conductor *alone* present, what is the field "above" and "below" the plate? **Key 48**
20. Finally, superimpose these fields with those found previously (Key 46), taking direction into account.
21. In the absence of the test charge, what is the flux through a small sphere centered at P, according to Gauss' law? **Key 47**
22. If the electric field at some region of the sphere points toward P, what is the sign of the flux through that region? **Key 40**

Learning Guide 13   Electricity

ANSWER KEY

1. 7.5×10^{-6} —i.e., about one in 133,000 electrons must be removed.
2. 8×10^4 N/C, away from point A
3. $\sim 1.2 \times 10^{-13}$ J
4. $\Phi_E = d^{5/2} a(\sqrt{2} - 1)$
5. Small $m_e \Rightarrow$ too small E; how would you see electrons?
6. 9×10^{11} N, or $\sim 10^8$ tons, which is the weight of about 1500 battleships
7. 4×10^{-3} N
8. 8×10^{23} electrons
9. $F_y = -\dfrac{1}{4\pi\epsilon_0} \dfrac{2q^2}{a^2}$
10. $(F_x, F_y) = \dfrac{1}{4\pi\epsilon_0} \dfrac{q^2}{a^2}(4 + 1/\sqrt{2}, -2 + 1/\sqrt{2})$.
    Resultant: $|\mathbf{F}| = 0.176$ N.
    Direction: $\theta = 15.3°$ clockwise from $2q \to -2q$ line.
11. About 1.2×10^{-13} J
12. $E_1 = Q_1/2A\epsilon_0$; $E_2 = E_3 = -E_1$
13. $\mathbf{E} = \dfrac{Q}{4\pi\epsilon_0 l y}(-\mathbf{i} + \mathbf{j})$
14. $U \simeq \tfrac{1}{2}\langle mv^2 \rangle \simeq \tfrac{3}{2}kT$ for each molecule.
15. 4×10^{-3} N
16. 7.32×10^{-19} C
17. 6×10^{18} electrons
18. $E = F/2q$, where F is given in Key 10.
19. $T \simeq 10^{10}$ K
20. No. Gauss' law requires only that
    $$\oint \mathbf{E} \cdot d\mathbf{A} = 0$$
    Yes, the converse is true.

21. $E_x = -\dfrac{Q}{4\pi\epsilon_0 l}\left(\dfrac{1}{y} - \dfrac{1}{\sqrt{y^2 + l^2}}\right),$
    $E_y = \dfrac{Q}{4\pi\epsilon_0 l y} \sin\theta_0.$
22. 1.64×10^{-19} C
23. 2×10^4 N/C, away from point B
24. No
25. No (if the charge is inside the cube)
26. Newton's third law requires $\mathbf{F}_{AB} = -\mathbf{F}_{BA}$.
27. A concentric spherical surface of radius r, with $r < R_1$ for part (a), $R_1 < r < R_2$ for part (b), and $r > R_2$ for part (c).
28. $\Phi_E = \oiint \mathbf{E} \cdot d\mathbf{A} = 0$
    for each of these four walls.
29. $\Phi_E = \oiint \mathbf{E} \cdot d\mathbf{A} = 0.$
    $\mathbf{E} = 0$ everywhere on the surface because the surface lies inside a conductor through which no current is flowing.
30. $\Phi_E = 0$
31. Yes; $\Phi_E = 0$.
32. $\Phi_E = |\mathbf{E}| A \cos\alpha$
33. For the wall at $x = d$,
    $$\int \mathbf{E} \cdot d\mathbf{A} = -(a\sqrt{d})d^2;$$
    and for the wall at $x = 2d$,
    $$\int \mathbf{E} \cdot d\mathbf{A} = +(a\sqrt{2d})d^2.$$
34. $Q_{\text{enc}} = 0$
35. $Q_{\text{inside}} = \epsilon_0 d^{5/2} a(\sqrt{2} - 1)$
36. $E_r = (q_1 + q_2)/4\pi\epsilon_0 r^2$ for $r > R_2$
37. Even though charges may separate, the total charge on the conductor of radius R_1 must be q_1,

and the total charge on the conductor of radius R_2 must be q_2.

38. No, the flux through the surface depends only on the charge inside the surface.

39. Shell of radius R_1: q_1 uniformly distributed on outer surface.

Shell of radius R_1: $-q_1$ uniformly distributed on inner surface; $q_1 + q_2$ uniformly distributed on outer surface.

40. Negative

41. $\Phi_E = \oiint \mathbf{E} \cdot d\mathbf{A} = E_r 4\pi r^2$

42. (i) $\sigma = -Q_1/2A$; (ii) $\sigma = Q_1/2A$

43. $E_1 = \dfrac{Q_1 + Q_2}{2A\epsilon_0}$;

$E_2 = \dfrac{Q_2 - Q_1}{2A\epsilon_0}$;

$E_3 = -E_1$

44. $Q_{\text{enclosed}} = \begin{cases} 0 & \text{for (1)} \\ q_1 & \text{for (2)} \\ q_1 + q_2 & \text{for (3)} \end{cases}$

45. $E_r = q_1/4\pi\epsilon_0 r^2$ for $R_1 < r < R_2$.

46. $E_1 = Q_1/2A\epsilon_0$; $E_2 = E_3 = -E_1$

47. Zero

48. $E_2 = Q_2/2A\epsilon_0 = E_1$; $E_3 = -E_2$

49. From Gauss' law
$$\Phi_E = \oiint \mathbf{E} \cdot d\mathbf{A} = \dfrac{1}{\epsilon_0} Q_{\text{inside}}.$$

50. No (if the charge is inside the cube)

51. $\mathbf{E}$ is in the radial direction, and $|\mathbf{E}|$ is the same for all directions at a given r.

52. Yes

53. $E_r = 0$ for $r < R_1$.

learning guide 14

Electric Potential

Suggested Reading: Fishbane, Gasiorowicz, & Thornton, Chapters 25 and 26

PROBLEM I

The following questions test your knowledge of electric potentials. If you have any difficulty, reread Chapter 25.

1. Do electrons tend to go to regions of *low potential* or *high potential*?
 Key 53
2. Suppose that the earth has a net charge that is not zero. Can one still adopt the earth as a standard reference point of potential and assign the potential $V = 0$ to it? **Key 7**
3. What scheme can you devise to ensure that the electric potential in a given region of space will have a constant value? **Key 16**

PROBLEM II

Consider a point charge with $q = 3.0 \times 10^{-8}$ C.

1. What is the radius of an equipotential surface having a potential of 60 V?
 Key 28
2. Are surfaces whose potentials differ by a constant amount (say, 1.0 V) evenly spaced in radius? **Key 57**

If your answer is incorrect, try Helping Question 1.

3. What work must be done against electric forces to move an electron from a rest position at some point on the 60-V equipotential surface to a rest position infinitely far from the stationary 3.0×10^{-8} C charge? Express your answer in electron-volts (eV) and in joules (J). See Helping Question 2 if needed. **Key 51**

PROBLEM III

An idealized device for accelerating electrons can be made as shown in the diagram. Two metal plates are placed in a vacuum and connected to a voltage supply that puts negative charges on the lower plate and positive charges on the upper plate until there is a potential difference of 300 V between the plates. An electron, which may be "boiled off" a piece of hot metal not shown, starts from (nearly) rest at the bottom plate. It is accelerated upward until it passes through a hole in the top plate.

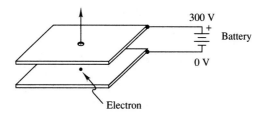

How fast is the electron going when it passes through the hole? (Recall that $e = -1.6 \times 10^{-19}$ C and $m_e = 9.11 \times 10^{-31}$ kg.) If you have trouble, see Helping Questions 2, 3, and 4. **Key 60**

PROBLEM IV

A thin glass rod of length $l = 1.0$ m is rubbed with a furry object until it has a uniformly distributed charge, $Q = 10^{-9}$ C. What is the electric potential at a perpendicular distance $s = 0.5$ m from its end? Helping Questions 5 through 9 may be needed. **Key 36**

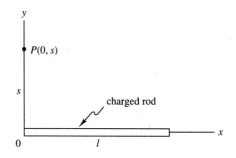

Learning Guide 14   Electric Potential

PROBLEM V

A quartet of point charges $+e, -e, +e, -e$ is arranged at the corners of a square of side $a = 1$ Å (1 Å $= 10^{-10}$ m). Find the potential energy of this charge configuration, relative to the case in which $a = \infty$. See Helping Questions 10 through 14 if you have trouble.   **Key 56**

Now, suppose we have nine such arrangements of point charges, themselves arrayed in a square as shown, such that the point charges are all the same distance a away from their nearest neighbors. Note that in this arrangement, each point charge is attracted to its (four) nearest neighbors, so that the whole array tends to hold itself together. This sort of arrangement, extended into three dimensions, is the basis of crystals of ionic compounds such as NaCl and LiF. Such crystals are held together primarily by the electrostatic attractions among their constituent ions.

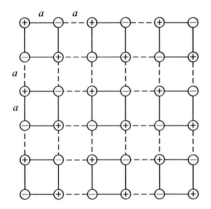

PROBLEM VI

A charge Q and a charge $5Q$ are located on the x-axis as shown in the diagram. The fields produced by these charges are observed at a point with coordinates (X, Y, Z).

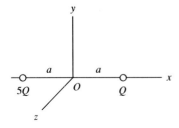

1. Find the electric field at the point (X, Y, Z) by using Coulomb's law. ($E_x = ?$ $E_y = ?$ $E_z = ?$) See Helping Questions 15 and 16 if you have trouble, and Helping Question 17 if you are still having trouble. **Key 1**
2. Find the potential V at the point (X, Y, Z) by using the $1/r$ law. If you have trouble, see Helping Questions 15 and 18. **Key 55**
3. Find the electric field at the point (X, Y, Z) by differentiating the potential you found in part (2). If you don't remember which equation to use, see Fishbane, Gasiorowicz, & Thornton, Eq. (25-29). **Key 8**
4. Find the potential difference between a point $(0, Y, 0)$ on the y-axis and the origin of the coordinate system $(0, 0, 0)$ by integrating $-\mathbf{E} \cdot d\mathbf{l}$ over a straight-line path between these two points. Help in setting up the integral can be found in Helping Questions 19 to 23. **Key 52**
5. Compute $V(0, Y, 0) - V(0, 0, 0)$ using the results of part (2) and compare your result in part (4). Do they agree? **Key 14**
6. If you explicitly evaluated $V(0, Y, 0) - V(0, 0, 0)$ by integrating $-\mathbf{E} \cdot d\mathbf{l}$ along a different path, would you get the same answer? **Key 47**

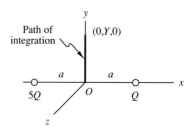

PROBLEM VII

A parallel-plate capacitor has capacitance C_0 and plate separation d. Two dielectric slabs, of constants κ_1 and κ_2, each of thickness $d/2$, are inserted between the plates. Charges Q and $-Q$ are put on the upper and lower plates.

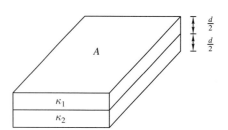

1. Without the dielectrics, what is the field $\mathbf{E}_0$ between the plates (in terms of Q, C_0, and d)? **Key 4**
2. *With* the dielectrics, what are the fields in dielectric 1 and dielectric 2? Stuck? See Helping Question 24. **Key 48**
3. What is the potential difference between the two plates? If you need to, see Helping Question 25. **Key 10**
4. What is the capacitance? **Key 23**

Learning Guide 14   Electric Potential

PROBLEM VIII

A parallel-plate capacitor has plates of area A and separation d and is charged to a potential difference V. The charging battery is then disconnected, and the plates are pulled apart until their separation is $3d$. Derive expressions in terms of A, d, and V for

1. The new potential difference. **Key 39**
2. The initial and the final stored energy. **Key 50**
3. The work required to separate the plates. **Key 17**

If you have trouble, see Helping Questions 26 and 27.

PROBLEM IX

A potential difference of 500 V is applied to a 2.0-μF capacitor and an 8.0-μF capacitor connected in series.

1. What are the charge and the potential difference for each capacitor?
Key 34
2. The charged capacitors are reconnected with their positive plates together and their negative plates together, no external voltage being applied. What are the charge and the potential difference for each? **Key 11**
3. The charged capacitors in part (1) are reconnected with plates of opposite sign together. What are the charge and the potential difference for each?
Key 25

If you have trouble, Helping Question 28 should help.

PROBLEM X

A sphere of radius R is filled with a *uniform* volume charge density and has total charge Q.

1. Find the electrostatic energy density at distance r from the center of the sphere for $r < R$ and $r > R$. If help is needed, see Helping Question 29.
Key 3
2. Compute the total electrostatic energy. **Key 19**
3. Explain why the value obtained in part (2) is greater than that for a spherical conductor of radius R carrying a total charge Q. **Key 44**

PROBLEM XI

When uranium ^{235}U captures a neutron, it splits into two nuclei (and emits several neutrons that can cause other uranium nuclei to split). Assume that the fission products are two equally charged nuclei with charge $+46e$ and that these nuclei are essentially at rest just after fission and separated by twice their radius $2R = 1.3 \times 10^{-14}$ m.

1. Calculate the electrostatic potential energy of the fission products. This is approximately the energy released per fission. If troubled, try Helping Question 30. **Key 30**
2. How many fissions per second are needed to produce 1 MW of power in a reactor? **Key 5**
3. How much ^{235}U will be consumed in a year in such a reactor? **Key 9**

HELPING QUESTIONS

1. What is the potential at a distance r from a charge q? **Key 46**
2. How is the change in the electron's potential energy related to the change in potential? **Key 12**
3. What principle can be used to provide the answer immediately? **Key 41**
4. What equation expresses this principle for the case at hand? **Key 45**
5. What is the expression for the potential due to a continuous charge distribution? If you are wrong, review Fishbane, Gasiorowicz, & Thornton, Section 25-5. **Key 18**
6. Let λ be the charge per meter on the rod. What is the charge dq on an element dx of the rod? If you don't get the right answer, see Fishbane, Gasiorowicz, & Thornton, Section 24-3. **Key 54**
7. In terms of s and the coordinate x of an element of charge, what is the distance of the element of the charge from P? **Key 2**
8. Express the answer to the problem as an integral over x, using the ends of the rod as limits. **Key 38**
9. Using a table of integrals, obtain the result.
10. With only charge 1 in place, what is the electric potential at corner 2? If you are wrong, review Fishbane, Gasiorowicz, & Thornton, Section 25-2. **Key 22**
11. As charge 2 is brought from ∞ to corner 2, what is the change in its potential energy? If you don't get the correct answer, reread Fishbane, Gasiorowicz, & Thornton, Section 25-2. **Key 27**
12. With charges 1 and 2 in place, what is the electric potential at corner 3? If your answer is wrong, review Fishbane, Gasiorowicz, & Thornton- Section 25-5. **Key 15**
13. Continue the procedure begun in Helping Questions 10 through 12. What is the next step? **Key 58**
14. What is the sum of the changes in potential energy required to assemble all four charges?

    <u>Thought question</u>: Call the value of the ith charge Q_i ($i = 1, 2, 3, 4$) and let $\mathbf{r}_i$ be its final position. By imagining the charges to be assembled in the order 1, 2, 3, 4, you obtained

$$V_{\text{total}} = \frac{1}{4\pi\epsilon_0}\left[\frac{Q_1 Q_2}{|\mathbf{r}_1 - \mathbf{r}_2|}\right.$$
$$+ \left(\frac{Q_1 Q_3}{|\mathbf{r}_1 - \mathbf{r}_2|} + \frac{Q_2 Q_3}{|\mathbf{r}_2 - \mathbf{r}_3|}\right)$$
$$+ \left(\frac{Q_1 Q_4}{|\mathbf{r}_1 - \mathbf{r}_4|}\right.$$
$$\left.\left.+ \frac{Q_2 Q_4}{|\mathbf{r}_2 - \mathbf{r}_4|} + \frac{Q_3 Q_4}{|\mathbf{r}_3 - \mathbf{r}_4|}\right)\right].$$

Can you write this expression in a way that makes it apparent that the order in which the charges were assembled is irrelevant? **Key 6**

15. What is the distance between the observation point (X, Y, Z) and the position $(-a, 0, 0)$ of the charge $5Q$? **Key 26**

16. What are the components (n_x, n_y, n_z) of the vector $\hat{\mathbf{n}}$ of unit length that points from the charge $5Q$ to the observation point? This vector gives the field $\mathbf{E}$ produced at the observation point by the charge $5Q$:

$$\mathbf{E}_{5Q} = \frac{5Q}{4\pi\epsilon_0} \frac{1}{r^2} \hat{\mathbf{n}}.$$

Key 59

17. The electric field produced by several charges is equal to the _____ of the fields produced by each charge separately. **Key 40**

18. The potential produced by several charges is equal to the _____ of the potentials produced by each charge separately. **Key 37**

19. Let R denote the distance from the origin to a point $(0, Y, 0)$ on the path. Then we can write the displacement vector $d\mathbf{l}$ from one point to another along the path in the form $d\mathbf{l} = \mathbf{n}dR$. What is $\mathbf{n}$? **Key 49**

20. Write the integral in the form

$$\Delta V = \int_0^Y (\text{something})\,dR,$$

where "something" is written in terms of $\mathbf{E}$ and $\mathbf{n}$. **Key 21**

21. Using your explicit expression for $\mathbf{E}$, write out the integral for ΔV. **Key 33**

22. What substitution will enable you to do the integral easily? **Key 24**

23. What is the resulting integral? **Key 20**

24. What is the effect of introducing a dielectric of constant κ in a region of field $\mathbf{E}_0$? **Key 31**

25. What is the potential difference across the dielectric κ_1? **Key 42**

26. Does the charge change when the plates are moved? **Key 32**

27. What is the new value of the capacitance? **Key 43**

28. What do you know about the charges on capacitors connected in series, and in parallel? **Key 13**

29. What is the energy density u in an electric field? **Key 29**

30. What is the electrostatic potential energy of two charges, q_1 and q_2, separated by a distance $2R$? **Key 35**

ANSWER KEY

1. $E_x = $
$$\frac{5Q}{4\pi\epsilon_0} \frac{X+a}{[(X+a)^2 + Y^2 + Z^2]^{3/2}}$$
$$+\frac{Q}{4\pi\epsilon_0} \frac{X-a}{[(X-a)^2 + Y^2 + Z^2]^{3/2}};$$

$E_y = $
$$\frac{5Q}{4\pi\epsilon_0} \frac{Y}{[(X+a)^2 + Y^2 + Z^2]^{3/2}}$$
$$+\frac{Q}{4\pi\epsilon_0} \frac{Y}{[(X-a)^2 + Y^2 + Z^2]^{3/2}};$$

$$E_z = \frac{5Q}{4\pi\epsilon_0} \frac{Z}{[(X+a)^2 + Y^2 + Z^2]^{3/2}}$$
$$+ \frac{Q}{4\pi\epsilon_0} \frac{Z}{[(X-a)^2 + Y^2 + Z^2]^{3/2}}$$

2. $r = (s^2 + x^2)^{1/2}$

3. $u = \dfrac{Q^2}{32\pi^2\epsilon_0 r^4}$, for $r > R$;

   $u = \dfrac{Q^2 r^2}{32\pi^2\epsilon_0 R^6}$, for $r < R$.

4. $|\mathbf{E_0}| = Q/C_0 d$

5. 2.67×10^{16} fissions per second.

6. $V_{\text{total}} = \dfrac{1}{2} \sum_{i \neq j}^{i,j} \dfrac{1}{4\pi\epsilon_0} \dfrac{Q_i Q_j}{|\mathbf{r}_i - \mathbf{r}_j|}.$

   In doing the sum one must count each pair (i, j) only once—that's why there's a factor of $\frac{1}{2}$ in front.

7. Yes, as long as the potential difference between two points on the surface is negligible.

8. Should be the same as key 1.

9. 328 g

10. $V = \dfrac{Q}{2C_0}\left(\dfrac{1}{\kappa_1} + \dfrac{1}{\kappa_2}\right)$

11. $Q_2 = 3.2 \times 10^{-4}$ C; $Q_8 = 12.8 \times 10^{-4}$ C; $\Delta V_2 = \Delta V_8 = 160$ V.

12. $\Delta U = -e\,\Delta V$, where e is the electron charge.

13. Capacitors connected in series have equal charges. When connected in parallel the potential difference across the capacitors is the same for each capacitor.

14. Yes, hopefully

15. $V_3 = \dfrac{e}{4\pi\epsilon_0 a}\left(\dfrac{1}{\sqrt{2}} - 1\right)$

16. Put a conducting surface around the (empty) region.

17. $W = \epsilon_0 A V^2/d$

18. $V = \dfrac{1}{4\pi\epsilon_0} \displaystyle\int \dfrac{dq}{r}$

19. $U = \dfrac{3Q^2}{20\pi\epsilon_0} \dfrac{1}{R}$

20. $\Delta V = -\dfrac{6Q}{4\pi\epsilon_0} \displaystyle\int_{|a|}^{\sqrt{a^2+y^2}} \dfrac{dr}{r^2}$

21. $\Delta V = -\displaystyle\int_0^Y dR\,[E_x(0, R, 0)n_x]$
    $\quad - \displaystyle\int_0^Y dR\,[E_y(0, R, 0)n_y]$
    $\quad - \displaystyle\int_0^Y dR\,[E_z(0, R, 0)n_z]$
    $= -\dfrac{6Q}{4\pi\epsilon_0} \displaystyle\int_0^Y \dfrac{R\,dR}{(a^2 + R^2)^{3/2}}$

22. $V_2 = e/4\pi\epsilon_0 a$

23. $C = C_0 \dfrac{2\kappa_1 \kappa_2}{(\kappa_1 + \kappa_2)}$

24. $r = \sqrt{a^2 + R^2}$; $dr = R\,dR/r$

25. $Q_2 = Q_8 = 0$; $\Delta V_2 = \Delta V_8 = 0$.

26. $r = \sqrt{(X+a)^2 + Y^2 + Z^2}$

27. $\Delta U_2 = -e^2/4\pi\epsilon_0 a$

28. 4.5 m

29. $u = \frac{1}{2}\epsilon_0 E^2$

30. $U = k\dfrac{(46)^2 e^2}{2R} = 3.75 \times 10^{-11}$ J

31. $E_0 \to E_0/\kappa$

32. No

33. $\Delta V = -\dfrac{6Q}{4\pi\epsilon_0} \displaystyle\int_0^Y \dfrac{R\,dR}{(a^2 + R^2)^{3/2}}$

34. $Q_2 = Q_8 = 8.0 \times 10^{-4}$ C; $\Delta V_2 = 400$ V; $\Delta V_8 = 100$ V.

35. $U = kq_1 q_2/2R$

36. $V = \dfrac{Q}{4\pi\epsilon_0 l}\left[\ln\left(\dfrac{x}{s} + \sqrt{1 + \dfrac{x^2}{s^2}}\right)\right]_0^l$
    $= 13$ V

37. Sum

38. $V = \dfrac{\lambda}{4\pi\epsilon_0} \displaystyle\int dx\,(x^2 + s^2)^{-1/2}$

39. $V_{\text{new}} = 3V$

40. Vector sum

41. Conservation of energy

42. $\Delta V = \text{field} \times \text{distance} = \dfrac{E_0 d}{\kappa_1 2}$

43. $C_{new} = \dfrac{\epsilon_0 A}{3d}$; $\dfrac{1}{3}$ of C_{old}

44. For a spherical conducting shell of radius R, all of the charge Q resides on the conductor—i.e., at $r = R$. The electric field inside such a shell is zero, but for $r > R$ the electric field is the same as that for the uniform volume distribution. Hence the total electrostatic energy is less for the conducting shell.

45. $\tfrac{1}{2}mv^2 - e\,\Delta V = 0$

46. $V = q/4\pi\epsilon_0 r$

47. You'd better believe it!

48. $E_1 = Q/C_0 d\kappa_1$, $E_2 = Q/C_0 d\kappa_2$

49. **n** = unit vector in the direction of the path. $n_x = 0$; $n_y = 1$; $n_z = 0$.

50. Initial energy = $\dfrac{1}{2}\left(\dfrac{\epsilon_0 A}{d}\right)V^2$

    Final energy = $\dfrac{3}{2}\left(\dfrac{\epsilon_0 A}{d}\right)V^2$

51. $W = 60$ eV $= 9.6 \times 10^{-18}$ J

52. $\Delta V = V(0, Y, 0) - V(0, 0, 0)$

    $= \dfrac{6Q}{4\pi\epsilon_0}\left(\dfrac{1}{\sqrt{a^2 + Y^2}} - \dfrac{1}{|a|}\right)$

53. High potential because their charge is negative.

54. $dq = \lambda\,dx = (10^{-9}\,\text{C/m})\,dx$

55. $V = $

    $\dfrac{5Q}{4\pi\epsilon_0}\dfrac{1}{\sqrt{(X+a)^2 + Y^2 + Z^2}}$

    $+ \dfrac{Q}{4\pi\epsilon_0}\dfrac{1}{\sqrt{(X-a)^2 + Y^2 + Z^2}}$

56. $U = \dfrac{e^2}{4\pi\epsilon_0 a}\left[-1 + \left(-1 + \dfrac{1}{\sqrt{2}}\right)\right.$

    $\left. - \left(2 - \dfrac{1}{\sqrt{2}}\right)\right]$

    $= -5.9 \times 10^{-18}$ J

57. No

58. Multiply the answer to Helping Question 12 by $+e$ to get the change in potential energy required to bring charge 3 into position.

59. $n_x = (X + a)/r$; $n_y = Y/r$; $n_z = Z/r$

60. $v = 1.0 \times 10^7$ m/s, which is 1/30 the speed of light.

learning guide 15

Currents and DC Circuits

Suggested Reading: Fishbane, Gasiorowicz, & Thornton, Chapters 27 and 28

PROBLEM I

A ring of radius R with charge per unit length λ rotates with angular velocity ω about its axis. Find an expression for the current I at a point on the ring. See Helping Questions 1 through 3 for help. **Key 4**

PROBLEM II

A current of 15 A exists in #12 copper wire, which has a diameter of 2 mm. (Copper has mass density $\rho_m = 8.92$ g/cm^3 and resistivity $\rho = 1.7 \times 10^{-8}\ \Omega \cdot$m.)

1. What is the current density J? **Key 64**
2. What is the drift velocity of the electrons (assume one free electron per Cu atom)? If help is needed, review Fishbane, Gasiorowicz, & Thornton, Section 27-4. **Key 45**
3. What is the average time between collisions of the electrons with the lattice ions? **Key 34**

4. Assuming an average speed of 2×10^6 m/s for the electrons, what is their mean free path in this wire? **Key 53**

If you do not know how to do parts (3) and/or (4), read carefully Fishbane, Gasiorowicz, & Thornton, Section 27-4, and Examples 27-5 and 27-6.

PROBLEM III

A solid rectangular material has edges of length $a, b,$ and c, where $a < b < c$.

1. Across which pair of opposite faces will a constant electric potential give the maximum current? Stuck? Try Helping Question 4. **Key 38**
2. It is desired to measure the resistivity of a solid rectangular piece of moon rock that has dimensions $a = 2$ cm, $b = 3$ cm, and $c = 5$ cm. A 10-V potential is applied to the opposite bc faces, and a current of 6.0 μA is measured. What is the resistivity? **Key 20**
3. Would you expect to measure the same current if you applied the 10-V potential to either of the two remaining opposite faces? Why? **Key 30**

PROBLEM IV

A 1000-Ω resistor is rated at 2 W.

1. What is the maximum current allowed through this resistor? If your answer is incorrect, see Fishbane, Gasiorowicz, & Thornton, Section 27-7. **Key 43**
2. If this maximum current exists in the 1-kΩ (1000-Ω) resistor for 5 minutes, how many coulombs pass through the resistor in this time? **Key 16**
3. How many electrons pass through any cross section of the resistor in this time? **Key 11**

PROBLEM V

A resistor consists of a cylindrical shell of length L with inner and outer radii r_1 and r_2, respectively. With a constant potential difference $(V_2 - V_1)$ between inner and outer cylindrical surfaces, find the resistance in terms of $L, r_1, r_2,$ and ρ, the resistivity. **Key 2**

If you are correct, go to Problem VI; if you have trouble, use as many of Helping Questions 5 through 9 as you need to.

Learning Guide 15 Currents and DC Circuits

PROBLEM VI

Given the circuits below:

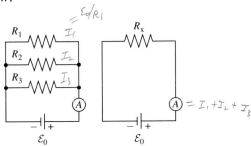

Annotations on figure: $= \mathcal{E}_0/R_1$, I_1, I_2, I_3, $\text{(A)} = I_1 + I_2 + I_3$

1. Find the current flowing through each of the resistors R_1, R_2, R_3. For help, see Helping Question 10. **Key 36**

2. What current is read by the ideal ammeter A in circuit 1? For help, see Helping Question 11. $A = \mathcal{E}_0/R_x \quad \frac{1}{R_x} = \frac{1}{R_1} + \frac{1}{R_2} + \frac{1}{R_3}$ **Key 14**

3. Circuit 2 is equivalent to circuit 1; i.e., the same battery emf $\mathcal{E}_0$ is present and the same current flows through the two ammeters. Find the equivalent resistance R_x in terms of R_1, R_2, R_3. See Helping Question 12 for help. **Key 7**

4. Find the ratio I_1/I_{total} of the current through R_1 to the current through the ammeter in terms of R_x and R_1. See Helping Question 13 if you need assistance. **Key 13**

$$\frac{I_1}{I_{\text{total}}} = \frac{R_x}{R_1}, \quad \frac{I_2}{I_{\text{total}}} = \frac{R_x}{R_2}$$

PROBLEM VII

Consider the circuit shown in the diagram. In this circuit, R_i is the resistance of the battery (that is, it cannot be removed and has to be considered as a part of the battery); R_0 is a fixed resistance that is larger than R_i; R_V is a variable resistance, which is not initially connected; and A and B are the battery terminals.

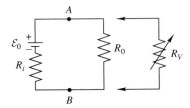

1. Explain what is meant by the emf of the battery. If you don't know, reread Fishbane, Gasiorowicz, & Thornton, Section 28-1. **Key 61**

2. What is the effective voltage that would be measured between A and B when R_0 is connected as shown? See Helping Question 14 for assistance. **Key 21**

3. What amount of power, appearing as heat, is dissipated in the resistance R_0? **Key 48**

4. An experimentalist wishes to extract the greatest amount of external power possible from the battery. He connects the additional resistance R_V in the manner shown by the arrows and varies it until the total heat generated in R_0 and R_V together is a maximum. What is the value of R_V required, and what is the value of the combined parallel resistance of R_V and R_0? **Key 37**

If your answer is correct, you have discovered the principle of "impedance matching." Congratulations! Go on to Problem VIII. Otherwise, see Helping Questions 15 and 16.

PROBLEM VIII

In the accompanying circuit, the switch S is closed at time $t = 0$ (C is uncharged at $t = 0$).

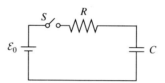

1. What is the current I_R through the resistor at time $t = 0$? **Key 6**
2. What is I_R at time $t \to \infty$? **Key 42**
3. Using the loop theorem, write the differential equation for the charge $q(t)$ on the capacitor that describes this circuit. **Key 17**
4. In general, the solution to a differential equation of the form

$$\frac{dy}{dt} + ay + b = 0$$

is $y = Ae^{pt} + B$, where A, B, and p are constants to be determined. Substitute this general solution into the equation you found in part (3). Group constant terms and terms involving Ae^{pt}. What conditions must be imposed on A, B, and p to obtain a solution that satisfies the differential equation? **Key 40**
5. In order to determine the one remaining constant, what condition must be used? (See the first sentence of this problem.) **Key 41**
6. After the switch is closed, what is the charge on the plates of C at any time? **Key 28**
7. How much energy is stored in the capacitor? **Key 68**
8. How much energy has been dissipated in the resistance R by time t_1? See Helping Questions 17 through 19. **Key 3**
9. Now plug some numbers into your answers for parts (1), (6), (7), and (8). Use $\mathcal{E} = 5$ V, $R = 1000$ Ω, and $C = 1.0$ μF. For parts (6), (7), and (8), use $t_1 = 10^{-3}$ s. **Key 1**

Learning Guide 15 Currents and DC Circuits

PROBLEM IX

1. Write an equation relating I_1, I_2, and I_3. **Key 46**
2. Write expressions for the voltage drops around the two loops (*abcd*) and (*befc*) in terms of given voltages, resistances, and currents. If you have problems, see Fishbane, Gasiorowicz, & Thornton, Section 28-3. **Key 66**

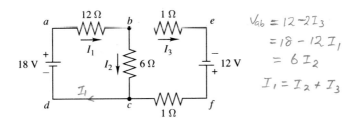

3. Solve for currents I_1, I_2, and I_3. **Key 32**
4. What is the significance of the signs of the currents? **Key 69**

PROBLEM X

Two concentric spherical metal shells of negligible thickness have a substance of dielectric constant κ and resistivity ρ between them. The radius of the inner shell is a, and of the outer shell b. A battery of constant emf V_0 is connected between the inner and outer shells. Assume that charges distribute themselves uniformly over the spheres and that the current flow is spherically symmetric. (The metallic surfaces are perfectly conducting.)

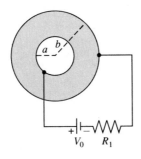

1. What are the capacitance C and the resistance R of this device? See Helping Questions 20 through 23 for assistance. **Key 52**
2. Draw an equivalent circuit in terms of R, C, V_0, and R_1, the internal battery resistance. **Key 59**
3. Find the current through the battery after it has been connected for a time that is long compared with the time constant of this device. If you didn't get the right answer, try Helping Question 24. **Key 67**
4. Find the charge q on the outer shell as a function of time, assuming the battery was connected at time $t = 0$. If troubled, see Helping Questions 25 through 30. **Key 39**

5. What is the time constant τ_c of the device? **Key 54**

6. After a long time compared with τ_c, the battery is disconnected. Find the charge q on the outer shell T seconds after disconnection. If incorrect, see Helping Questions 31 and 32. **Key 19**

HELPING QUESTIONS

1. If a charge Δq flows through an area in time Δt, what is the electric current? If your answer doesn't match, read Fishbane, Gasiorowicz, & Thornton, Section 27-1, over again. **Key 63**

2. What is the charge Δq contained in length ΔL of the ring? **Key 51**

3. In time Δt what charge Δq will flow through a point on the ring? **Key 58**

4. How is current related to geometry? If correct, retry Problem III, part (1). **Key 33**

5. For a given current I, what is the current density $\mathbf{J}$ at some distance r from the center? **Key 10**

6. If you don't see this, consider a concentric cylindrical surface of radius r, length L. How much current is going through it? What is its area? Now try Helping Question 5 again. **Key 5**

7. Now get $\mathbf{E}$ as a function of r from $\mathbf{J}(r)$. **Key 22**

8. Express $V_2 - V_1$ as a function of E. **Key 29**

9. You should be able to get $V_2 - V_1$ as a function of I now. You just have to know the definition of resistance to finish the problem. **Key 27**

10. What voltage appears across each resistor? **Key 18**

11. Does charge conservation help? **Key 9**

12. Ohm's law! **Key 23**

13. Use Ohm's law again or combine the answers from parts (1), (2), and (3).

14. Apply the loop theorem to this loop. **Key 26**

15. Regard the combined resistance of R_0 and R_V as a new resistance R_e, use the expression from Problem VII, part (3), and vary this resistance (now R_e instead of R_0) to find a maximum in the value of the heat in R_e. **Key 24**

16. What value of R_V in parallel with R_0 makes the combination equal to the value of R_e that you found in Helping Question 15? **Key 35**

17. How does the current vary with time? **Key 12**

18. How is the energy dissipated in the resistance R related to the current? **Key 8**

19. What is the value of $\int_0^{t_1} e^{-\alpha t}\, dt$? **Key 31**

20. The capacitance of the device is the ratio of the charge q to the potential difference $V_a - V_b$. What is $V_a - V_b$ for a spherical capacitor? **Key 57**

21. The resistance of the device is the ratio of the voltage $(V_a - V_b)$ to the current I flowing from sphere a to sphere b. First, what is the current density J when a current I flows through a sphere of radius r, where $(a < r < b)$? **Key 47**

22. Second, using Ohm's law, find the electric field E at a radius

Learning Guide 15   Currents and DC Circuits

r, where $a < r < b$. If you don't get the correct answer, review Fishbane, Gasiorowicz, & Thornton, Section 27-3. **Key 55**

23. Third, what is $V_a - V_b$? If you don't get this, see Fishbane, Gasiorowicz, & Thornton, Section 25-5. **Key 62**

24. When the capacitor is charged, the current I flowing through the battery goes entirely through R. What does the loop theorem give for the loop (V_0–R_1–R–V_0)? If your answer is incorrect, review Fishbane, Gasiorowicz, & Thornton, Section 28-2. **Key 49**

25. Let I denote the current through the battery, I_C the current through C, and I_R the current through R. What relation does the junction theorem give? Wrong answer? Read Fishbane, Gasiorowicz, & Thornton, Section 28-3, over again. **Key 56**

26. If q denotes the charge on the capacitor, what is I_C? If you have trouble, review Fishbane, Gasiorowicz, & Thornton, Section 28-5. **Key 65**

27. What does the loop theorem yield for I_R? **Key 60**

28. What does the loop theorem give for the loop (V_0–R_1–C–V_0)? **Key 50**

29. If, in the answer to Helping Question 28, I is replaced by its value in terms of q, what relation does one obtain? **Key 15**

30. Use the method outlined in Problem VIII, parts (4) and (5), to determine the solution to the differential equation obtained in Helping Question 29.

31. For $t \gg \tau_c$, what is q? **Key 25**

32. When the battery is disconnected, what is the time constant τ_c' of the discharging circuit? If you don't get the right answer, have another look at Fishbane, Gasiorowicz, & Thornton, Section 28-5. **Key 44**

ANSWER KEY

1. (1) 5 mA; (6) 3.16×10^{-6} C; (7) 4.96×10^{-6} J; (8) 10.8×10^{-6} J.

2. $R = \dfrac{\rho}{2\pi L} \ln \dfrac{r_2}{r_1}$

3. $E_{\text{diss}} = \tfrac{1}{2}\mathcal{E}^2 C \left(1 - e^{-2t/RC}\right)$

4. $I = \lambda \omega R$

5. I; $2\pi r L$

6. $I_R = \mathcal{E}/R$

7. $\dfrac{1}{R_x} = \dfrac{1}{R_1} + \dfrac{1}{R_2} + \dfrac{1}{R_3}$

8. Heat dissipated = $\int_0^{t_1} I^2(t) R \, dt$.

9. The current through ammeter A comes from the branches containing R_1, R_2, and R_3.

10. $|\mathbf{J}| = I/2\pi r L$; $\mathbf{J}$ is radially outward.

11. 8.3×10^{19} electrons

12. $I = \dfrac{dq}{dt} = \dfrac{\mathcal{E}}{R} e^{-t/RC}$

13. $I_1/I_{\text{total}} = R_x/R_1$

14. $I_{\text{total}} = I_1 + I_2 + I_3$
    $= \mathcal{E}_0 \left(\dfrac{1}{R_1} + \dfrac{1}{R_2} + \dfrac{1}{R_3}\right)$

15. $\dfrac{dq}{dt} + \dfrac{q}{C}\left(\dfrac{1}{R} + \dfrac{1}{R_1}\right) = \dfrac{V_0}{R_1}$

16. 13.4 C

17. $\mathcal{E} = IR + q/C$; $I = dq/dt$ is the clockwise current; q is the charge on C.
18. $\mathcal{E}_0$
19. $q(T) = \dfrac{CV_0 R}{R + R_1} e^{-T/RC}$
20. 1.25×10^5 Ω·m
21. $\mathcal{E}_{\text{eff}} = \mathcal{E} R_0/(R_0 + R_i)$
22. $|\mathbf{E}| = \rho|\mathbf{J}| = \rho I/2\pi r L$
23. $\mathcal{E}_0 = I_{\text{total}} R_x$
24. $R_e = R_i$
25. $q(t \gg t_c) = CV_0 R/(R + R_1)$
26. $\mathcal{E} - IR_i - IR_0 = 0$
27. $V_2 - V_1 = I \dfrac{\rho}{2\pi L} \ln \dfrac{r_2}{r_1}$
28. $q(t) = \mathcal{E} C (1 - e^{-t/RC})$
29. $V_2 - V_1 = -\int_{r_1}^{r_2} \mathbf{E} \cdot d\mathbf{r}$
30. No. The resistance depends on the geometry.
31. $\int_0^{t_1} e^{-\alpha t}\, dt = \dfrac{1}{\alpha}\left(1 - e^{-\alpha t_1}\right)$
32. $I_1 = 2$ A, $I_2 = -1$ A, $I_3 = 3$ A
33. I is proportional to area and to (thickness)$^{-1}$.
34. $\tau = 2.5 \times 10^{-14}$ s
35. $R_V = R_0 R_e/(R_0 - R_e)$
36. $I_1 = \mathcal{E}_0/R_1$; $I_2 = \mathcal{E}_0/R_2$; $I_3 = \mathcal{E}_0/R_3$
37. $R_V = R_0 R_i/(R_0 - R_i)$; combined resistance $= R_i$.
38. Across the bc faces
39. $q(t) = \dfrac{CV_0 R}{R + R_1}$
   $\times \left[1 - e^{-\frac{t}{C}\left(\frac{1}{R} + \frac{1}{R_1}\right)}\right]$
40. $p + (1/RC) = 0$; $(B/RC) - (\mathcal{E}/R) = 0$; A not determined.
41. At $t = 0$, $q(t) = 0$, because C is not charged.
42. $I_R \to 0$ as $t \to \infty$.
43. 44.7 mA
44. $\tau'_c = RC$
45. $v_d = J/ne = 3.6 \times 10^{-4}$ m/s
46. $I_1 = I_2 + I_3$
47. $J = I/4\pi r^2$
48. Power $= I^2 R_0 = \dfrac{\mathcal{E}^2}{(R_0 + R_i)^2} R_0$
49. $IR_1 + IR = V_0$
50. $V_0 = IR_1 + q/C$
51. $\Delta q = \lambda \Delta L$
52. $C = 4\pi \mathcal{E}_0 \kappa ab/(b-a)$; $R = \rho(b-a)/4\pi ab$
53. $l = 5 \times 10^{-8}$ m
54. $\tau_c = \dfrac{C}{(1/R + 1/R_1)}$
55. $E = \rho J = \rho I/4\pi r^2$
56. $I = I_C + I_R$
57. $V_a - V_b = \dfrac{-q}{4\pi \mathcal{E}_0 \kappa} \int_b^a \dfrac{dr}{r^2}$
   $= \dfrac{q}{4\pi \mathcal{E}_0 \kappa}\left(\dfrac{1}{a} - \dfrac{1}{b}\right)$
58. $\Delta q = \lambda \omega R \, \Delta t$
59.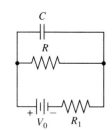
60. $I_R = \dfrac{1}{R}\dfrac{q}{C}$
61. Potential difference across battery when there is no current
62. $V_a - V_b = -\int_b^a E\, dr$
   $= \dfrac{\rho I}{4\pi}\left(\dfrac{1}{a} - \dfrac{1}{b}\right)$
63. $I = \dfrac{\Delta q}{dt}$
64. $J = 4.8 \times 10^6$ A/m^2
65. $I_C = dq/dt$
66. $-12I_1 - 6I_2 + 18 = 0$; $-I_3 + 12 - I_3 + 6I_2 = 0$.
67. $I = V_0/(R + R_1)$
68. $q^2/2C$
69. I_1 and I_3 flow in the directions of the arrows; I_2 flows opposite to the arrow.

learning guide 16

Magnetism

Suggested Reading: Fishbane, Gasiorowicz, & Thornton, Chapters 29 and 30

PROBLEM I: AN EXERCISE WITH VECTOR CROSS PRODUCTS

In the following parts, you are asked to evaluate certain vector cross products. The answers to parts (1), (2), and (4) are to be expressed in terms of the unit vectors **i**, **j**, and **k**.

1. Evaluate **a** × **b** for the diagram shown. What is **b** × **a**?   Key 9

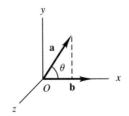

2. Evaluate **a** × **b**. Key 19

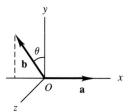

3. In what plane (xy, xz, or yz) does the vector **a** × **b** lie? What is the angle between this vector and the y-axis? Key 3

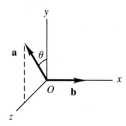

4. Suppose $\mathbf{a} = 5\mathbf{i} - \mathbf{j}$ and $\mathbf{b} = \mathbf{j} + \mathbf{k}$. What is **a** × **b**? If you can't do this part, look at Helping Question 1. If necessary, Helping Question 2 will probably show you the rest of what you need. Key 17

PROBLEM II

Suppose that a region of space contains uniform electric and magnetic fields of unknown direction and magnitude. In order to measure these fields, a test charge of $Q = 10^{-6}$ C is used. (In practice, there are much better ways to measure the fields, but we want to illustrate how the basic definitions work.) The following measurements are made:

1. At rest, the test charge experiences a force of $\mathbf{F} = 10 \times 10^{-5}$ N in the direction of the positive y-axis.
2. Moving at a speed of $v = 20$ m/s in the positive y-direction, the charge experiences a force given by

$$\mathbf{F} = (10\mathbf{j} - 2\mathbf{k}) \times 10^{-5} \text{ N}.$$

3. If the charge moves with the same speed in the positive z-direction, the force is

$$\mathbf{F} = (-\mathbf{i} + 12\mathbf{j}) \times 10^{-5} \text{ N}.$$

Use this information to find the electric and magnetic fields. If you don't see the solution, refer to as many of Helping Questions 3 to 5 as you may need. Key 15

Learning Guide 16   Magnetism

PROBLEM III

A uniform magnetic field of magnitude 1 T points vertically from up to down. At a certain time, a 1-MeV proton is moving horizontally from east to west through the field.

1. What is the direction of the force acting on the proton at this time? If you don't get the right answer, review Fishbane, Gasiorowicz, & Thornton, Section 29-2.   **Key 7**
2. What is the magnitude of the force acting on the proton? If you have trouble with this, try Helping Question 6, and, if necessary, 7.   **Key 12**
3. The proton will subsequently move in a circle. How many revolutions will the proton perform per second? If your answer is incorrect, review Fishbane, Gasiorowicz, & Thornton, Section 29-3.   **Key 22**
4. Derive an equation for the radius of the circle in terms of the proton's momentum p, its charge q, and the magnetic field B. What is the radius for this case? Puzzled? See Fishbane, Gasiorowicz, & Thornton, Section 29-3.   **Key 1**
5. What is the kinetic energy of the proton after it has completed one quarter of a revolution? If you don't see the answer, refer to Helping Question 8.   **Key 21**

PROBLEM IV

A positive charge q is uniformly distributed around the edge of a record of radius a and mass M. It is rotating as it would on a record player, with angular velocity ω downward and normal to the plane of the record.

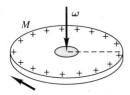

1. What is the magnetic moment of the record? See Helping Questions 9 and 10 for assistance.   **Key 6**
2. Suppose that a uniform magnetic field **B** directed into the page is turned on. There will be a torque τ on the record. Find its magnitude and direction. If you don't see how to do this, use Helping Questions 11 and 12 and/or review Fishbane, Gasiorowicz, & Thornton, Section 29-5.   **Key 10**

PROBLEM V

Show that it is impossible to design a magnet that produces a uniform magnetic field in the space between the magnet's poles and zero field outside of this space.

(See the figure. In real magnets, the magnetic field drops to zero in a continuous fashion, as shown in the figure.)

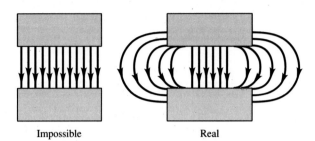

Impossible   Real

If you have convinced yourself, go to Problem VI. If you're not sure, see Helping Questions 13 to 15.

PROBLEM VI

A long wire carries a current I_1. The rectangular loop carries a current I_2. Calculate the resulting force acting on the loop due to the current I_1. If after a good try you are still puzzled, see Helping Question 16. If after that you are still in the dark, try Helping Questions 17 and 18. **Key 43**

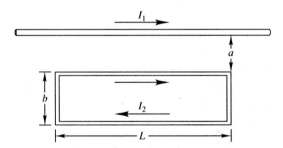

PROBLEM VII

A conductor consists of an infinite number of adjacent wires, each infinitely long and carrying a current I. Show that the lines of **B** will be represented as shown in the figure. If there are n wires per unit length, what is the magnitude of **B**? **Key 37**

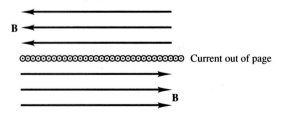

Learning Guide 16   Magnetism

PROBLEM VIII

A current loop in the form of a square with sides of length $2L$ carries a current I. What is the magnetic field at the center of the square? If your answer is correct, go to Problem IX. Otherwise, see Helping Questions 19 through 21.   **Key 50**

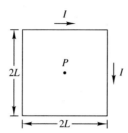

PROBLEM IX: HELMHOLTZ COILS

Two N-turn coils are arranged a distance apart equal to their radius, as in the figure. The coils carry a current I.

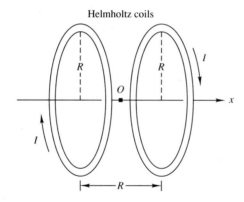

Helmholtz coils

1. What is the magnetic field $B(x)$ along the axis? (Take $x = 0$ at the center.) If you are having trouble, see Helping Questions 22 and 23.   **Key 33**
2. What are

$$B_x, \quad \frac{dB_x}{dx}, \quad \text{and} \quad \frac{d^2 B_x}{dx^2}$$

at $x = 0$? What do they indicate? See Helping Question 24 for assistance.
    Key 42

PROBLEM X

A plastic disk of radius R has a positive charge q uniformly distributed over its surface. If the disk is rotated at an angular frequency ω about its axis, find:

1. The **B** field at the center of the disk.　　　　　　　　　　　　**Key 32**
2. The magnetic dipole moment of the disk.　　　　　　　　　　　　**Key 48**

(**Hint:** The rotating disk is equivalent to an array of current loops; see Section 29-5 in Fishbane, Gasiorowicz, & Thornton.)

If after an honest effort you cannot get part (1), Helping Questions 25 through 27 should help. If you cannot get part (2), try Helping Questions 28 and 29.

PROBLEM XI

A circular copper loop of radius $a = 10$ cm carries a current of $I_1 = 30$ A. At its center is placed a second loop of radius $b = 1.0$ cm, having $N = 50$ turns and a current of $I_2 = 1.0$ A.

1. What magnetic field **B** does the large loop set up at its center? For help on this, see Fishbane, Gasiorowicz, & Thornton, Example 30-5.　　**Key 25**
2. What torque acts on the small loop? Assume that the planes of the two loops are at right angles and that the **B** field provided by the large loop is essentially uniform throughout the volume occupied by the small loop. If you're stuck, see Fishbane, Gasiorowicz, & Thornton, Section 29-5, and/or Helping Questions 30 and 31.　　**Key 44**

PROBLEM XII

Calculate the **B** field at a point P on the axis of a tightly wound solenoid of length L and radius a that has n turns of wire per unit length and carries a current I, assuming P to be *outside* the solenoid at a distance y from the end of the coil. If you need help, see Helping Questions 32 through 34.　　**Key 39**

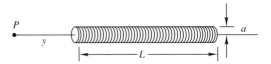

PROBLEM XIII

(Optional)

Given the solenoid of Problem XII and given a point P on its axis a distance $y < L$ from one end, calculate the **B** field at P, and show that it can be expressed as

$$B = \left(\frac{\mu_0 R I}{2}\right)[\cos\theta_1 + \cos\theta_2],$$

with θ_1 and θ_2 as shown in the figure.

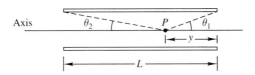

(**Hint:** This is essentially the same problem as Problem XII.)

HELPING QUESTIONS

1. How can $\mathbf{a} \times (\mathbf{b} + \mathbf{c})$ be expressed in terms of $\mathbf{a} \times \mathbf{b}$ and $\mathbf{a} \times \mathbf{c}$? **Key 13**

2. What are $\mathbf{i} \times \mathbf{j}$, $\mathbf{j} \times \mathbf{j}$, $\mathbf{j} \times \mathbf{k}$, and $\mathbf{i} \times \mathbf{k}$? **Key 20**

3. How is the total force on the charge expressed in terms of $\mathbf{E}$, $\mathbf{B}$, Q, and $\mathbf{v}$? **Key 4**

4. Using measurement (1), determine the electric field. **Key 14**

5. For measurements (2) and (3), express the force that the test charge should experience for an arbitrary B_x, B_y, and B_z. **Key 8**

6. What is the force $\mathbf{F}$ acting on a particle of charge q and velocity $\mathbf{v}$ moving in a magnetic field $\mathbf{B}$? **Key 18**

7. Check your units carefully. Have you converted from MeV to joules?

8. What work is done by the magnetic force on the proton? **Key 16**

9. For the definition of magnetic moment, see Fishbane, Gasiorowicz, & Thornton, Section 29-5. How much charge moves per rotational period through a small surface cutting the edge of the record? **Key 2**

10. What is the current due to this charge? **Key 23**

11. How is ω_p related to the torque τ and the angular momentum $\mathbf{L}$? **Key 5**

12. Find $\mathbf{L}$ in terms of ω, M, and a, assuming the record is a uniform disk. **Key 11**

13. Write down Ampère's law for a judiciously chosen path. A judiciously chosen path is shown in **Key 54**. Now retry Problem V. If you still have trouble, see Helping Questions 14 and 15.

14. Assuming, for the moment, that the field really was like that in the figure, what would be the integral $\oint \mathbf{B} \cdot d\mathbf{l}$ around the path? **Key 29**

15. What would be the current enclosed by the path, according to Ampère's law? What is the *actual* current enclosed by the path? **Key 38**

16. What is the magnitude of $\mathbf{B}$ at a distance r from a long, straight wire carrying a current I? **Key 47**

17. How does the force due to current I_1 on the short side of the rectangle on the left compare with the force on the short side on the right? **Key 34**

18. What is the force on the top side of the rectangle? **Key 53**

19. Consider each side of the square separately, as in the figure below.

What is the magnitude of **B** created at P by a length dx of the wire as shown? **Key 45**

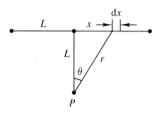

Now retry Problem VIII; if you still have difficulties, proceed with Helping Question 20.

20. Write an integral for the total B created at P by this wire, putting everything in terms of *either* x or θ. **Key 28**
21. In terms of θ, the integral is easy. If you do it in terms of x, you may need the result

$$\int \frac{dx}{(x^2 + a^2)^{3/2}} = \frac{x}{a^2(x^2 + a^2)^{1/2}}.$$

Either way should give the same answer. **Key 49**
22. See Fishbane, Gasiorowicz, & Thornton, Example 30-5, for the field due to one coil. What is the direction of the field $\mathbf{B}(x)$ due to the left-hand coil? What is the direction of the field $\mathbf{B}(x)$ due to the right-hand coil? **Key 30**
23. What is the distance x'_1 of the observation point x from the left-hand coil? What is the distance x'_2 of the observation point at x from the right-hand coil? **Key 41**
24. What is dB_x/dx? **Key 51**
25. What is the charge density per unit area? **Key 26**
26. What is the value of B at the center of a loop of radius r carrying current I? **Key 35**
27. What is the current dI due to the rotation of the disk for an annulus of radius r ($r < R$) and width dr? **Key 46**
28. What is the definition of magnetic moment? **Key 31**
29. What is the magnetic moment $d\mu$ of a circular strip of radius r and width dr? **Key 36**
30. What is the torque τ on a magnetic dipole? **Key 40**
31. What is the magnetic dipole moment μ? **Key 52**
32. Since the solenoid is tightly wound, one can consider it the equivalent of a line of rings of current (see Fishbane, Gasiorowicz, & Thornton, Section 30-4).
33. You may be off by a factor of n if you forget that the current in each ring of thickness dx is not I, but $nI\,dx$.
34. What are the limits of integration? The method of integration you should know. Check Helping Question 21. **Key 27**

ANSWER KEY

1. $r = p/qB = 14.5$ cm
2. q
3. yz-plane; angle of $(\pi/2 - \theta)$
4. $\mathbf{F} = Q(\mathbf{E} + \mathbf{v} \times \mathbf{B})$
5. $\boldsymbol{\tau} = \boldsymbol{\omega} \times \mathbf{L}$
6. $\mu = qa^2\omega/2$, downward.
7. South
8. For measurement (2):

$$\mathbf{F} = [10^{-4}\mathbf{j} + 2\times 10^{-5}(\mathbf{j}\times\mathbf{B})]\ \text{N}$$
$$= [10^{-4}\mathbf{j} + 2\times 10^{-5}(B_z\mathbf{i} - B_x\mathbf{k})]\ \text{N}.$$

For measurement (3):

$$\mathbf{F} = [10^{-4}\mathbf{j} + 2\times 10^{-5}(\mathbf{k}\times\mathbf{B})]\ \text{N}$$
$$= [10^{-4}\mathbf{j} + 2\times 10^{-5}(B_x\mathbf{j} - B_y\mathbf{i})]\ \text{N}.$$

9. $\mathbf{a}\times\mathbf{b} = (-ab\sin\theta)\mathbf{k}$; $\mathbf{b}\times\mathbf{a} = -\mathbf{a}\times\mathbf{b}$
10. $\tau = qa^2\omega B/2$, to the right.
11. $\mathbf{L} = Ma^2\omega/2$, downward.
12. 2.22×10^{-12} N
13. $\mathbf{a}\times(\mathbf{b}+\mathbf{c}) = (\mathbf{a}\times\mathbf{b}) + (\mathbf{a}\times\mathbf{c})$
14. $\mathbf{E} = 10^2\mathbf{j}$ N/C
15. $\mathbf{E} = 10^2\mathbf{j}$ N/C; $\mathbf{B} = (\mathbf{i}+\mathbf{j}/2)$ T
16. None
17. $\mathbf{a}\times\mathbf{b} = -\mathbf{i} - 5\mathbf{j} + 5\mathbf{k}$
18. $\mathbf{F} = q(\mathbf{v}\times\mathbf{B})$
19. $\mathbf{a}\times\mathbf{b} = (ab\cos\theta)\mathbf{k}$
20. $\mathbf{i}\times\mathbf{j} = \mathbf{k}$; $\mathbf{j}\times\mathbf{j} = 0$; $\mathbf{j}\times\mathbf{k} = \mathbf{i}$; $\mathbf{i}\times\mathbf{k} = -\mathbf{j}$.
21. 1 MeV
22. 1.52×10^7 rev/s
23. $I = q\omega/2\pi$
24. $\omega_p = qB/M$, out of page (axis rotates counterclockwise in plane of page).
25. $B = \mu_0 I_1/2a = 1.88\times 10^{-4}$ T, along the axis.
26. $\sigma = q/\pi R^2$
27. $x = y$ and $x = y + L$, or

$$\theta = \tan^{-1}\frac{a}{y}$$

and

$$\theta = \tan^{-1}\frac{a}{y+L}.$$

(The second set will of course depend on the exact substitution you use to solve the integral.)

28. $B = \dfrac{\mu_0 I}{4\pi}\displaystyle\int_{-L}^{L}\dfrac{\cos\theta\,dx}{r^2}$

$$= \dfrac{\mu_0 I L}{4\pi}\int_{-L}^{L}\dfrac{dx}{(x^2+L^2)^{3/2}}$$

$$= \dfrac{\mu_0 I}{4\pi L}\int_{-\pi/4}^{\pi/4}\cos\theta\,d\theta$$

29. $\oint \mathbf{B}\cdot d\mathbf{l} = Bh + 0 + 0 + 0$
30. $B_1(x)$ and $B_2(x)$ both point to the right, along the axis.
31. $\mu = IA$, where A is the loop area.
32. $B = \mu_0\omega q/2\pi R$, in the direction of ω.
33. $B_x = \dfrac{\mu_0 N I R^2}{2}$
$$\left\{\dfrac{1}{\left[R^2+\left(\tfrac{1}{2}R+x\right)^2\right]^{3/2}} + \dfrac{1}{\left[R^2+\left(\tfrac{1}{2}R-x\right)^2\right]^{3/2}}\right\}$$

34. Equal in magnitude but opposite in direction
35. $B = \mu_0 I/2r$
36. $d\mu = \dfrac{q\omega r^3\,dr}{R^2}$
37. $|\mathbf{B}| = \mu_0 n I/2$
38. $I = Bh/\mu_0$, $I_{\text{actual}} = 0$
39. $\mathbf{B} = \dfrac{\mu_0 n I}{2}\left[\dfrac{y+L}{\sqrt{a^2+(y+L)^2}} - \dfrac{y}{\sqrt{a^2+y^2}}\right]$,

along the axis.

40. $\boldsymbol{\tau} = \boldsymbol{\mu}\times\mathbf{B}$
41. $x_1 = R/2 + x$; $x_2 = R/2 - x$
42. $B_x(0) = \dfrac{8\mu_0 I N}{5\sqrt{5}R}$; $\dfrac{dB_x}{dx}(0) = 0$;

$$\frac{d^2B}{dx^2}(0) = 0.$$

The fact that

$$\frac{dB}{dx} = \frac{d^2B}{dx^2} = 0$$

at $x = 0$ means that B is *nearly uniform* near $x = 0$. Recall that for small x

$$B(x) = B(0) + x\frac{dB}{dx}(0) + \frac{x^2}{2}\frac{d^2B}{dx^2}(0).$$

43. $F = \dfrac{\mu_0 I_1 I_2 L}{2\pi}\left[\dfrac{1}{a} - \dfrac{1}{(a+b)}\right]$

(attractive).

44. $\tau = \mu_0 \pi N I_1 I_2 b^2 / 2a \simeq 3.0 \times 10^{-6}$ N·m

45. $dB = \dfrac{\mu_0 I}{4\pi}\dfrac{\cos\theta\, dx}{r^2}$,

into the paper.

46. $dI = d\left(\dfrac{dq}{dt}\right) = \dfrac{q\omega r\, dr}{\pi R^2}$

47. $B = \mu_0 I / 2\pi r$
48. $\mu = \omega q R^2 / 4$
49. $\mathbf{B} = \mu_0 I / (2\sqrt{2}\pi L)$, into the paper (contribution from one side).
50. $\mathbf{B} = \sqrt{2}\mu_0 I / \pi L$, into the paper.
51. At $x = 0$,

$$\frac{dB_x}{dx} = \frac{N\mu_0 I R^2}{2}$$

$$\times \left\{ \frac{-3(R/2 + x)}{\left[R^2 + (R/2 + x)^2\right]^{5/2}} \right.$$

$$\left. + \frac{3(R/2 - x)}{\left[R^2 + (R/2 - x)^2\right]^{5/2}} \right\} = 0.$$

52. $\boldsymbol{\mu} = (NI_2 A)\mathbf{n} = (NI_2\pi b^2)\mathbf{n}$
53. $|\mathbf{F}| = \dfrac{\mu_0 I_1 I_2 L}{2\pi a}$ (attractive).
54.

learning guide 17

Induction and Magnetism & Matter

Suggested Reading: Fishbane, Gasiorowicz, & Thornton, Chapters 31, 32, and 33

PROBLEM I

A circular loop of wire 6 cm in diameter is placed with its normal making an angle of 30° with the direction of a uniform 5000-G magnetic field. The loop is turned so that its normal rotates about the field direction at the constant rate of 100 rev/min; the angle between the normal and the field direction (= 30°) remains unchanged during this process. What emf appears in the loop? **Key 25**

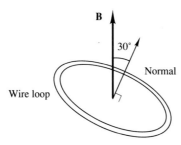

If your answer is correct, proceed to Problem II. Otherwise, use Helping Questions 1 and 2 sparingly.

PROBLEM II: ALTERNATING-CURRENT GENERATOR

A rectangular loop of N turns and of length a and width b is rotated at a frequency of f revolutions per second in a uniform magnetic field **B**, as shown in the figure. Let $t = 0$ at the instant shown.

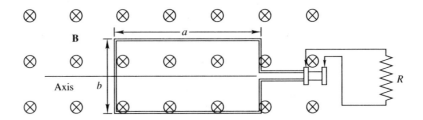

1. Calculate the magnitude of the induced emf that appears in the loop. This is the principle of the commercial alternating-current generator. **Key 40**

If you get the right answer, move on to part (2); otherwise use only as many of Helping Questions 3 through 5 as you need.

2. Design a loop that will produce a peak emf $\mathcal{E}_0 = 150$ V when rotated at 60 rev/s in a magnetic field of 5000 G. Express your answer by giving the required value of Nba. **Key 19**

PROBLEM III

A square wire of length ℓ, mass M, and resistance R slides without friction down parallel conducting rails of negligible resistance, as shown in the figure. The rails are connected to each other at the bottom by a resistanceless rail parallel to the wire, so that the wire and rails form a closed rectangular conducting loop. The plane of the rails makes an angle α with the horizontal, and a uniform vertical field of magnetic induction **B** exists throughout the region.

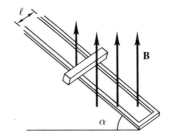

1. Find the magnitude of the steady-state speed acquired by the wire. **Key 41**

If your answer is correct, go on to part (2); otherwise, make use of as many of Helping Questions 6 through 11 as you need.

2. Prove that this result is consistent with the conservation-of-energy principle. An outline of the proof is given in Helping Questions 12 to 14. You may want to compare your proof with these questions.
3. What change, if any, would there be if B were directed down instead of up? **Key 5**

PROBLEM IV

1. Suppose that the flux of magnetic induction through the coil of N turns of the figure changes from Φ_1 to Φ_2. Find the charge q that flows through the circuit of total resistance R. **Key 42**

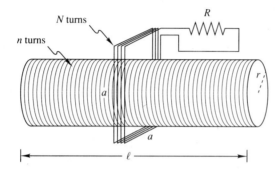

If your answer matches the key, go to part (2). If not, use Helping Questions 15 and 16 as you need.

2. Suppose the change in flux, $\Phi_2 - \Phi_1$, is zero. Does it follow that no Joule heating occurred during this time interval? **Key 39**
3. A solenoid of radius r and length ℓ is uniformly wound with n turns. A square coil of side a ($a \ll \ell$) and N turns is placed around it, near its center. If the current in the solenoid is changing at the rate of $dI/dt = \alpha$, what is the current in the resistance R? **Key 23**

If your answer is correct, go to part (4). If not, try Helping Question 17.

4. If the current in the solenoid changes from 0 to I_s, how much charge passes through the resistor R? (Solve this problem using part (3).) **Key 7**

If you get the right answer, move on to part (5). If not, try Helping Question 18.

5. Repeat part (4), this time using part (1) for the solution. If you need help, try Helping Question 19. **Key 7**

PROBLEM V

The figure shows a uniform magnetic field **B** confined to a cylindrical volume of radius R. The field **B** is increasing in magnitude at a constant rate of 100 G/s. What are the instantaneous accelerations (direction and magnitude) experienced by electrons of zero velocity placed at A, at O, and at C? Assume $r = 5.0$ cm. (The necessary fringing of the field beyond R will not change your answer as long as there is axial symmetry about a perpendicular axis through O.)

Key 33

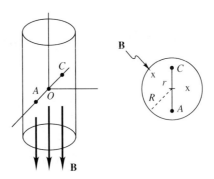

If you cannot work the problem, use as few of Helping Questions 20 to 22 as you can. If you have the right magnitude but the wrong sign, try Helping Question 23.

PROBLEM VI

In the old days, the magnetic field of the Princeton-Pennsylvania Accelerator (PPA) varied sinusoidally at 20 Hz (cycles/s) from 0.03 T to 1.39 T (1 T = 1 Wb/m^2). Suppose that the magnetic field has the same shape as in Problem V. If a silver dollar is placed at the center with its plane perpendicular to the field, how much power is converted into heat? A silver dollar has a radius $a = 2$ cm, a thickness $\ell = 2$ mm, and a resistivity $\rho = 1.6 \times 10^{-8}$ $\Omega \cdot$m.

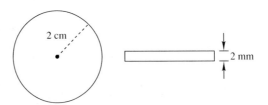

If you have trouble, Helping Questions 24 through 29 should help. Try, as always, to use as few as possible.

Key 35

PROBLEM VII

A coil with self-inductance L carries a current $I = I_0 \cos(2\pi f t)$. Find and graph, as functions of time:

1. The flux Φ_m.

Learning Guide 17 Induction and Magnetism & Matter 153

○—〰〰〰〰—○
      L
  →
  I

2. The self-induced emf $\mathcal{E}$. For help, see Helping Questions 30 and 31. **Key 51**

PROBLEM VIII

A circular coil of N_1 turns and area A_1 encircles a long, tightly wound solenoid of N_2 turns, area A_2, and length ℓ_2. The solenoid carries a current I_2.

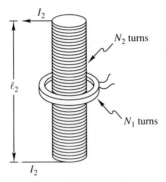

1. Find the flux Φ_{m1} through the circular coil due to the current I_2 in the solenoid. Does it depend on A_1? If help is needed, try Helping Questions 32 and 33. **Key 43**
2. Find the mutual inductance M of the solenoid and coil. Helping Question 34 may be useful. **Key 14**

PROBLEM IX

In the circuit below the switch S is closed at time $t = 0$.

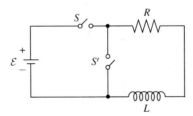

1. What is the current I through the circuit at time $t = 0$ and at time $t \to \infty$? **Key 54**
2. What is the voltage drop V_L across the inductance at $t = 0$ and at $t \to \infty$? **Key 47**

If help is needed, consult Fishbane, Gasiorowicz, & Thornton, Section 33-1.

After the switch S has been closed for a long time (i.e., a time long relative to the time constant of the circuit), it is opened and switch S' is closed. Let this instant be $t = 0$.

3. What is the current I at $t = 0$ and at $t \to \infty$? **Key 45**
4. What is the voltage drop V_R across the resistance at $t = 0$ and at $t \to \infty$? **Key 57**
5. Using the loop theorem, write the differential equation for the current $I(t)$. If you need help, reread Fishbane, Gasiorowicz, & Thornton, Section 33-4. **Key 49**
6. Find the current $I(t)$ by solving the differential equation you have just obtained. **Key 46**
7. What is the time constant of the circuit? **Key 8**
8. How much energy is stored in the inductance as a function of t? Section 33-2 of Fishbane, Gasiorowicz, & Thornton may be useful. **Key 56**
9. How much energy is dissipated in the resistance from $t = 0$ to $t = t_1$? If you experience difficulties, try Helping Questions 35 and 36. **Key 52**

HELPING QUESTIONS

1. What is an expression for the flux through the loop? **Key 38**
2. Does this expression depend on time? **Key 22**

   Retry Problem I. If you still have trouble, reread Fishbane, Gasiorowicz, & Thornton, Sections 31-2 and 31-3, paying particular attention to Eq. (31-2).

3. Let θ denote the angle between the normal to the plane of the loop and the **B** field. What is the flux through a single turn of wire as a function of θ? **Key 28**

   If you're having trouble, review Fishbane, Gasiorowicz, & Thornton, Section 24-1, for electric flux, which is analogous to magnetic flux in Section 31-2, and retry Helping Question 3.

4. What is the flux through N turns of the loop? **Key 10**
5. What is θ as a function of time? **Key 34**

   If your answer is correct, retry Problem II, part (1). If it isn't, review Fishbane, Gasiorowicz, & Thornton, Section 9-2, and retry Helping Question 5.

6. Let s denote the distance between the wire and the bottom of the rails, measured along the rails. What is the flux through the loop? **Key 1**

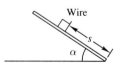

7. What is the induced emf in the loop when $ds/dt = v$? **Key 36**
8. How much current flows through the loop? **Key 20**
9. The magnetic force on the wire can be resolved into a component along the rails and a component perpendicular to the rails. What is the component along the rails? **Key 6**

Learning Guide 17 Induction and Magnetism & Matter

If you got the right answer, retry Problem III, part (1); otherwise, review Fishbane, Gasiorowicz, & Thornton, Section 29-4.

10. The gravitational force on the wire can also be resolved into components parallel and perpendicular to the rails. What is the component of the gravitational force along the rails? **Key 30**

    Retry Problem III, part (1).

11. How are the components found in Helping Questions 9 and 10 related when v is constant? **Key 17**

12. At what rate are gravitational forces doing work on the wire? **Key 37**

13. At what rate is energy being dissipated as heat? **Key 11**

    If you didn't get this, review Fishbane, Gasiorowicz, & Thornton, Section 27-7.

14. How are the energy rates of change in Helping Questions 12 and 13 related when v is constant? **Key 2**

15. Let Φ = flux through one turn of the coil. How is the current in the coil related to Φ? **Key 16**

    If you got the right answer to this, retry Problem IV, part (1). Or, if you are having problems, review Fishbane, Gasiorowicz, & Thornton, Sections 28-2 and 31-2.

16. How is the charge flowing through a circuit related to the current in the circuit? **Key 3**

    If you don't see the relation, review Fishbane, Gasiorowicz, & Thornton, Section 27-1. Otherwise, retry Problem IV, part (1).

17. If the current in the solenoid is I_s, what is the expression for the flux through the loop? **Key 31**

    If your answer is incorrect, review Fishbane, Gasiorowicz, & Thornton, Sections 30-3 and 31-2. If your answer is correct, retry Problem IV, part (3).

18. How is the charge related to the current? **Key 18**

    If you got the correct answer, retry Problem IV, part (4). If not, review Fishbane, Gasiorowicz, & Thornton, Section 27-1.

19. What are the initial and final fluxes through the loop? **Key 12**

    If your answer is the right one, retry Problem IV, part (5). If not, review Fishbane, Gasiorowicz, & Thornton, Sections 30-3 and 31-2.

20. What is the rate of change of magnetic flux through a circle of radius r? **Key 24**

21. What do the electric lines of force look like? **Key 32**

22. What is the electric field at a point a distance r from the axis? **Key 9**

23. What is the sign of
    (a) The magnetic flux Φ_B?
    (b) $d\Phi_B/dt$?
    (c) $\int \mathbf{E} \cdot d\mathbf{l}$ (integrated counter-clockwise)?
    (d) The charge of the electron? **Key 13**

24. Write an equation for the magnetic field as a function of time. **Key 26**

25. Consider an annular slice of the silver dollar with radius r and thickness dr. How much power is converted into heat in this slice? **Key 15**

    If your answer is correct, retry Problem VI. If you are unable to answer this helping question, go on to Helping Questions 26 through 29. Use them sparingly.

26. What is the resistance of this annular slice? **Key 29**

If you don't get this, reread Fishbane, Gasiorowicz, & Thornton, Section 27-3. Try to understand Eq. (27-16).

27. What is the emf around this ring? **Key 4**

28. What is the instantaneous power converted into heat in this ring? **Key 21**

If you need help with this, reread Fishbane, Gasiorowicz, & Thornton, Section 27-7.

29. What is the average value of $\cos^2(\omega t)$? **Key 27**

Retry Helping Question 25.

30. What is the definition of self-inductance? **Key 58**

If incorrect, read again Fishbane, Gasiorowicz, & Thornton, Section 33-1.

31. Apply Faraday's law.

32. What is the magnetic field B_2 inside the solenoid? **Key 48**

If incorrect, review Fishbane, Gasiorowicz, & Thornton, Section 30-3.

33. What is the flux of B_2 through the coil? **Key 55**

If your answer is incorrect, review Fishbane, Gasiorowicz, & Thornton, Section 31-2.

34. The flux Φ_{m1} is proportional to I_2. What is the name of the proportionality constant? **Key 44**

If you do not know, read Fishbane, Gasiorowicz, & Thornton, Section 33-1.

35. How much stored energy is lost in L from $t=0$ to $t=t_1$? **Key 50**

36. Where has the lost energy been dissipated? **Key 53**

ANSWER KEY

1. $\Phi_B = sBl\cos\alpha$

2. The two rates must be equal, since there are no other energy sources or sinks.

3. $I = dq/dt$

4. $\mathcal{E} = -\dfrac{d\Phi_B}{dt}$
   $= -\pi r^2 (2\pi f) B_0 \cos(2\pi f t)$

5. The current would flow in the opposite direction. The motion of the wire would be unchanged.

6. $\mathbf{F_B} = I\ell B\cos\alpha = vB^2\ell^2\dfrac{\cos^2\alpha}{R}$, pointing up the rails.

7. $q = Nn\mu_0 I_s \pi r^2 / \ell R$

8. L/R

9. $|\mathbf{E}| = \dfrac{r}{2}\left|\dfrac{d\mathbf{B}}{dt}\right|$.

   E points counterclockwise.

10. $NBba\cos\theta$

11. $I^2 R = v^2\ell^2 B^2 \dfrac{\cos^2\alpha}{R}$

12. $\Phi_1 = 0;\ N\Phi_2 = Nn\mu_0 I_s \pi r^2/\ell$

13. (a) −, (b) +, (c) −, (d) +

14. $M = \Phi_{m_1}/I_2 = \mu_0 N_1 N_2 A_2/\ell_2$

15. $dP = \dfrac{\pi^3 \ell f^2 B_0^2}{\rho} r^3 dr$

16. $I = \dfrac{N}{R}\dfrac{d\Phi}{dt}$

17. $\mathbf{F}_B = -\mathbf{F}_g$ (net force = 0).

18. $q_2 - q_1 = \displaystyle\int_1^2 I\, dt$

19. 0.796 m²·turn

20. $I = \mathcal{E}/R = B\ell v \cos\alpha/R$

21. $dP = \dfrac{\mathcal{E}^2}{R}$
$= \dfrac{2\pi^3 \ell f^2 B_0^2}{\rho} \cos^2(2\pi f t)\, r^3 dr$

22. No, although the loop wobbles, its radius doesn't change, the field doesn't change, and the angle between and normal to the loop and the field doesn't change.

23. $I = Nn\mu_0 \alpha \pi r^2/\ell R$

24. $\dfrac{d\Phi_B}{dt} = -\pi r^2 \dfrac{d|\mathbf{B}|}{dt}$

25. Zero

26. $B = B_{av} + B_0 \sin(2\pi f t)$,
$B_{av} = 0.71$ T,
$B_0 = 0.68$ T

27. $\langle \cos^2(\omega t)\rangle = \langle \sin^2(\omega t)\rangle = \tfrac{1}{2}$

28. $baB\cos\theta$

29. $R = \dfrac{2\pi r \rho}{\ell dr}$

30. $\mathbf{F}_g = Mg\sin\alpha$, pointing down the rails

31. $\Phi = Nn\mu_0 I_s \pi r^2/\ell$

32. Given the symmetry about the axis, the lines of force must be circles:

33. At A, $a = 4.4 \times 10^7$ m/s² to the left. At O, $a = 0$. At C, $a = 4.4 \times 10^7$ m/s² to the right.

34. $\theta = 2\pi f t$

35. $P = \dfrac{\pi^3 l R^4 f^2 B_0^2}{4\rho} = 29$ W

36. $\mathcal{E} = B\ell v \cos\alpha$

37. $\dfrac{dW_q}{dt} = \mathbf{F}_g \cdot \mathbf{v} = mgv\sin\alpha$

38. $\Phi_B = BA\cos\theta$
$= B\pi r^2 \cos\theta$
where B = magnetic field,
r = radius of loop,
and θ = angle between $\mathbf{B}$ and normal to loop.

39. No. Unless the flux remained constant, it would have induced some current flow. If $\Phi_1 - \Phi_2 = 0$, as much current flowed in the positive direction as in the negative direction. However, since the joule heating is given by $P = I^2 R$, the direction of current is irrelevant.

40. $\mathcal{E} = 2\pi f NabB \sin(2\pi f t) = \mathcal{E}_0 \sin(2\pi f t)$

41. $v = \dfrac{MgR\sin\alpha}{B^2 \ell^2 \cos^2\alpha}$

42. $q = N(\Phi_2 - \Phi_1)/R$

43. $\Phi_{m1} = \mu_0 N_1 N_2 A_2 I_2/\ell_2$; no, because the $\mathbf{B}$-field is zero outside a long, tightly wound solenoid.

44. The mutual inductance

45. $I(0) = \mathcal{E}/R;\ I(\infty) = 0$

46. $I(t) = \dfrac{\mathcal{E}}{R} e^{-Rt/L}$

47. $V_L(0) = \mathcal{E};\ V_L(\infty) = 0$

48. $B = \mu_0 N_2 I_2/\ell_2$

49. $L\dfrac{dI}{dt} + IR = 0$

50. $\Delta U_L = U_L(0) - U_L(t_1)$
$= \dfrac{L\mathcal{E}^2}{2R^2}\left(1 - e^{-2Rt_1/L}\right)$

51. (1) $\Phi_m = LI$
$= LI_0 \cos(2\pi f t)$
(2) $\mathcal{E} = -\dfrac{d\Phi_m}{dt}$
$= 2\pi f LI_0 \sin(2\pi f t)$

52. $U_R = \dfrac{L\mathcal{E}^2}{2R^2}\left(1 - e^{-2Rt_1/L}\right)$

53. In the resistance
54. $I(0) = 0;\quad I(\infty) = \mathcal{E}/R$
55. $\Phi = B_2 N_1 A_2$
56. $U_L = \dfrac{1}{2}LI^2 = \dfrac{L\mathcal{E}^2}{2R}e^{-2R_t/L}$
57. $V_R(0) = \mathcal{E};\quad V_R(\infty) = 0$
58. $L = \Phi/I$

learning guide 18

AC Circuits

Suggested Reading: Fishbane, Gasiorowicz, & Thornton, Chapter 34.

AC circuits may be analyzed using complex numbers or phasors. Complex numbers provide a much simpler and compact way to analyze circuits. In this Learning Guide we will use the complex number approach. See the supplemental notes at the end of this Learning Guide for additional details on this approach.

PROBLEM 1

In the circuit shown in the figure, switch S, initially in position a, is thrown to position b.

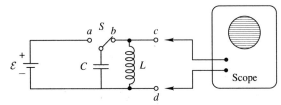

1. What waveform will be observed on the oscilloscope for the voltage between c and d? (Remember that the oscilloscope is a very "fast" voltmeter with practically infinite internal resistance.) **Key 1**

2. What is the period of the waveform? For help, see Fishbane, Gasiorowicz, & Thornton, Section 33-4. **Key 6**
3. What is the maximum charge on the capacitor? **Key 11**
4. What is the maximum current? Once again, Section 33-4 may be of use if you're having trouble. **Key 16**
5. When the charge Q on the capacitor is one-half the maximum charge, how much of the energy stored in the circuit is electrostatic and how much is magnetic? Helping Questions 1 through 3 may be needed. **Key 21**
6. What voltage is observed on the oscilloscope when the energy in the circuit is half electrostatic and half magnetic? Try Helping Questions 4 and 5 for assistance. **Key 26**
7. How can the frequency of the oscillating circuit be increased without changing the maximum current? If you do not understand the answer, try Helping Questions 6 and 7. **Key 31**

PROBLEM II

A resistor and a capacitor are connected in parallel across a sinusoidal emf, $\mathcal{E} = \mathcal{E}_{max} \cos(\omega t)$, as shown in the figure.

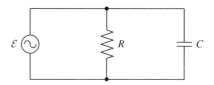

1. Find the current I_R in the resistor. **Key 36**

If your answer is correct, proceed to part (2). Otherwise, use Helping Question 8.

2. Find the current I_C through the capacitor. Helping Questions 9 and 10 may be needed. **Key 41**
3. Find the total current. Use only as many of Helping Questions 11 through 14 as needed. **Key 46**

PROBLEM III

Consider the *LCR* circuit shown in the figure, where R, L, and C are in parallel.

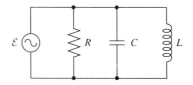

Learning Guide 18   AC Circuits

1. What are the currents flowing through the resistor, the capacitor, and the inductor? Use Helping Questions 15 through 17 as needed. **Key 2**
2. What is the total current? **Key 7**

If your answer isn't correct, see Helping Questions 18 through 21 and/or review Fishbane, Gasiorowicz, & Thornton, Sections 28-3 and 34-2.

3. Can any power be dissipated in the inductor or the capacitor? **Key 17**
4. It appears that the average power dissipated in this circuit is independent of frequency. Does this make sense? **Key 22**

PROBLEM IV

In the circuit shown in the figure, a resistor of resistance R is in series with a parallel LC circuit. In solving this problem, use what you have learned about series LCR circuits (Fishbane, Gasiorowicz, & Thornton, Section 33-4) and parallel LCR circuits (Problem III).

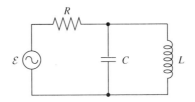

1. What is the equivalent impedance Z_{LC} of the parallel LC circuit? **Key 27**

If you do not know the answer or are puzzled by it, go back to the answer to part (2) of Problem III and let $R \to \infty$.

2. Why is there a minus sign in the expression for Z_{LC}? **Key 32**
3. Obtain the impedance Z of the circuit. If need be, use Helping Question 22. **Key 37**
4. What is the current flowing in this circuit? See Helping Questions 23 and 24 if you need to. **Key 42**
5. What is the time-averaged power dissipation in this circuit? Refer to Helping Questions 25 and 26 for assistance. **Key 12**
6. Make a sketch of the power dissipated as a function of frequency ω. What is the resonance condition? **Key 3**
7. Can you explain why no power is dissipated in the resistor when the circuit is driven at resonance? **Key 8**

HELPING QUESTIONS

1. How much energy is stored in the circuit? **Key 13**
2. If a charge Q is on the capacitor, how much electrostatic energy is stored in the capacitor? **Key 18**
3. If $Q = Q_{max}/2$, what fraction of the total energy is stored in the capacitor? **Key 23**
4. When $U_e = U_{total}/2$, what is Q/Q_{max}? **Key 28**
5. When a charge Q is on a capacitor, what is the voltage across it? **Key 33**
6. How can the frequency be increased? **Key 38**
7. Look at the answer to part (4), and the answer to part (7) will be obvious.
8. What does the loop theorem give for the loop formed by the voltage source and the resistance R? **Key 43**
9. What does the loop theorem give for the loop formed by the voltage source and the capacitance C? **Key 4**
10. What is the relation between current and charge? **Key 9**
11. What is the (complex) impedance of the parallel RC combination? **Key 14**
12. What is the (complex) current? **Key 19**
13. Rewrite the current found in the previous question as $I = I_{max}e^{i(\omega t + \phi)}$, where I_{max} is real. What is I_{max}? What is ϕ? **Key 24**
14. What, therefore, is the real part of the current? **Key 29**
15. What does the loop theorem give for the loop containing the resistor and the voltage source? **Key 34**
16. What does the loop theorem give for the loop containing the capacitor and the voltage source? **Key 39**
17. What does the loop theorem give for the loop containing the inductor and the voltage source? **Key 44**
18. What is the (complex) impedance of the parallel RLC combination? **Key 5**
19. What is the (complex) current? **Key 10**
20. Rewrite the current found in the previous question as $I = I_{max}e^{i(\omega t + \phi)}$, where I_{max} is real. What is I_{max}? What is ϕ? **Key 15**
21. What, then, is the real part of the current? **Key 20**
22. You have a resistor (impedance Z_R) in series with the LC circuit whose equivalent impedance is Z_{LC}. What formula would you use to find the equivalent impedance of the entire circuit? **Key 25**
23. Since you know the applied voltage and the equivalent impedance, what do you do to find the current? **Key 30**
24. Don't forget to find the amplitude and the phase angle as you did in Problems II and III. **Key 35**
25. Since you know the voltage and the current, how do you find the instantaneous power? **Key 40**
26. What is the time average of $\cos(\omega t)\cos(\omega t + \phi)$? **Key 45**

ANSWER KEY

1. A sinusoidal waveform, $\mathcal{E} = \mathcal{E}_0 \cos(\omega t)$

2. $I_R = \left(\dfrac{\mathcal{E}_{max}}{R}\right)\cos(\omega t)$,

   $I_C = (\omega C \mathcal{E}_{max})\cos(\omega t + 90°)$,

   and

   $I_L = \left(\dfrac{\mathcal{E}_{max}}{\omega L}\right)\cos(\omega t - 90°)$.

3. $\omega = \omega_0$

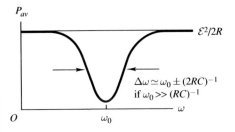

4. $\mathcal{E} = Q/C$, or $\mathcal{E} = IZ_C = I/(i\omega C)$.

5. $\dfrac{1}{Z} = \dfrac{1}{R} + i\omega C + \dfrac{1}{i\omega L}$

6. $T = 2\pi/\omega = 2\pi\sqrt{LC}$

7. $I = \dfrac{\mathcal{E}_{max}}{R}\sqrt{1 + \dfrac{R^2(\omega^2 LC - 1)^2}{(\omega^2 L^2)}}$

   $\times \cos(\omega t + \phi)$

   where

   $\phi = \tan^{-1}\left[\dfrac{R(\omega^2 LC - 1)}{(\omega L)}\right]$.

8. At resonance, oscillations in the LC part of the circuit keep the voltage at the "LC end" of the resistor the same as the voltage at the "$\mathcal{E}$ end" of the resistor, so no current flows through the resistor.

9. $I = i\omega Q$

10. $I = \mathcal{E}_{max}\left[\dfrac{1}{R} + i\dfrac{\omega^2 LC - 1}{\omega L}\right] e^{i\omega t}$

11. $C\mathcal{E}_0$

12. $P_{av} = \dfrac{1}{2}\dfrac{\mathcal{E}_{max}^2}{R}\dfrac{(\omega^2 - \omega_0^2)^2}{(\omega^2 - \omega_0^2)^2 + (\omega/RC)^2}$

13. $Q_{max}^2/2C$

14. $1/Z_{RC} = 1/R + i\omega C$

15. $I_{max} = \dfrac{\mathcal{E}_{max}}{R}\sqrt{1 + \dfrac{R^2(\omega^2 LC - 1)^2}{\omega^2 L^2}}$

    and

    $\phi = \tan^{-1}\left[\dfrac{R(\omega^2 LC - 1)}{\omega L}\right]$.

16. $I_{max} = \mathcal{E}_0 \sqrt{C/L}$

17. No—not in ideal inductors or capacitors

18. $Q^2/2C$

19. $I = [\mathcal{E}_{max}(1 + i\omega RC)/R]\, e^{i\omega t}$

20. $I = I_{max}\cos(\omega t + \phi)$, where I_{max} and ϕ are as in Helping Question 20.

21. $\dfrac{1}{4}, \dfrac{3}{4}$

22. Yes, the only dissipation is in the resistor, and the potential drop across the resistor is independent of frequency, since it is connected directly to the emf.

23. $\dfrac{1}{4}$

24. $I_{max} = \dfrac{\mathcal{E}}{R}\sqrt{(1 + \omega^2 R^2 C^2)}$

    and $\phi = \tan^{-1}(\omega RC)$.

25. $Z = Z_R + Z_{LC}$

26. $\mathcal{E}_0/\sqrt{2}$

27. $\dfrac{1}{Z_{LC}} = i\omega C + \dfrac{1}{i\omega L}$

    thus $Z_{LC} = \dfrac{i\omega L}{(1 - \omega^2 LC)}$

28. $1/\sqrt{2}$

29. $I = I_{max}\cos(\omega t + \phi)$, where I_{max} and ϕ are as in Helping Question 13.

30. $I = \mathcal{E}/Z$

31. By decreasing L and C in the same ratio
32. Because the currents through the inductor and capacitor are 180° out of phase
33. Q/C
34. $\mathcal{E} = IR$
35. I_{max}
$$= \frac{\mathcal{E}_{max}(\omega^2 - \omega_0^2)}{\sqrt{R^2(\omega^2 - \omega_0^2)^2 + (\omega^2/C^2)}};$$
$$\phi = \tan^{-1}\left[\frac{\omega/C}{R(\omega^2 - \omega_0^2)}\right].$$
36. $I_R = \mathcal{E}_{max}\cos(\omega t)/R$
37. $Z = \dfrac{R(\omega^2 LC - 1) - i\omega L}{\omega^2 LC - 1}$
$$= \frac{R(\omega^2 - \omega_0^2) - i\omega/C}{\omega^2 - \omega_0^2},$$
where $\omega_0 = 1/\sqrt{LC}$.
38. By decreasing L, C, or both
39. $\mathcal{E} = I/(i\omega C)$
40. $P_{inst} = \text{Re}\,[\mathcal{E}]\,\text{Re}[I]$
41. $I_C = \omega C \mathcal{E}_{max}\cos(\omega t + 90°)$

42. $I = \dfrac{\mathcal{E}_{max}(\omega^2 LC - 1)}{\sqrt{R^2(\omega^2 LC - 1)^2 + \omega^2 L^2}}$
$\cos(\omega t + \phi),$
where
$$\phi = \tan^{-1}\left[\frac{\omega L}{R(\omega^2 LC - 1)}\right].$$
This can also be written as
$$I = \frac{\mathcal{E}_{max}(\omega^2 - \omega_0^2)}{\sqrt{R^2(\omega^2 - \omega_0^2)^2 + (\omega^2/C^2)}}$$
$\cos(\omega t + \phi),$
where
$$\phi = \tan^{-1}\left[\frac{\omega/C}{R(\omega^2 - \omega_0^2)}\right].$$
43. $\mathcal{E} = IR$
44. $\mathcal{E} = i\omega LI$
45. $\tfrac{1}{2}\cos\phi$
46. $I = \dfrac{\mathcal{E}_{max}}{R}\sqrt{(1 + \omega^2 R^2 C^2)}$
$\cos(\omega t + \phi),$
where $\phi = \tan^{-1}(\omega RC).$

Notes: Linear Differential Equations, Complex Impedance, RLC Circuits, and Resonance

The purpose of these notes is to inform you about a very simple and elegant method that can be used to solve and understand many kinds of problems in physics.

RLC Circuits and Differential Equations

The starting point of this section is an RLC circuit such as the one shown in the figure (note that for a series arrangement the order of the parts around the loop doesn't affect the equations).

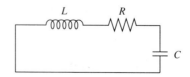

Learning Guide 18   AC Circuits

The principle of conservation of energy says that the rate of change of the stored electric and magnetic energy in the inductance and capacitance is equal to the rate of dissipation of energy in the resistance by Joule heating—i.e.,

$$\frac{d}{dt}\left(\frac{LI^2}{2} + \frac{Q^2}{2C}\right) = -I^2 R. \tag{1}$$

If you substitute $I = dQ/dt$ and cancel out I throughout, you get a voltage drop equation, which says that when you take a path around the circuit, the voltages across the various components must sum to zero:

$$L\frac{d^2 Q}{dt^2} + R\frac{dQ}{dt} + \frac{Q}{C} = 0. \tag{2}$$

This is a simple RCL circuit equation. The purpose of this supplement is to show you how to solve it.

One thing to note right away is that if you think of charge as a quantity like a position, then a current dQ/dt is like a velocity and $d^2 Q/dt^2$ is like an acceleration. Look at the analogous differential equation for damped simple harmonic motion, which is Eq. (13-52) in Fishbane, Gasiorowicz, & Thornton. You can see the analogies: m is like L; that is, an inductance is some sort of electromagnetic inertia. k is like $1/C$; that is, a capacitance is like a spring except that a small capacitance has a big spring constant. Ohmic resistance is like a gooey dashpot of viscous resistance.

Now, on to the mathematics.

THEOREM: Linear differential equations with constant coefficients, such as the following

$$a\frac{d^m}{dt^m}Q(t) + b\frac{d^{(m-1)}}{dt^{(m-1)}}Q(t)$$
$$+ \cdots + l\frac{d}{dt}Q(t) + mQ(t) = 0 \quad \text{(homogeneous)} \tag{3}$$

$$a\frac{d^m}{dt^m}Q(t) + b\frac{d^{(m-1)}}{dt^{(m-1)}}Q(t)$$
$$+ \cdots + l\frac{d}{dt}Q(t) + mQ(t) = K \quad \text{(inhomogeneous)} \tag{4}$$

can always be solved by solutions of the following sort:

$$Q(t) = Ae^{pt} \quad \text{(homogeneous)} \tag{5}$$

and

$$Q(t) = Ae^{pt} + B \quad \text{(inhomogeneous)}, \tag{6}$$

where A, B, and p are constants.

To prove that these are in fact the solutions just substitute them into equations and see that it all works out.

Let's try it on equation (2) right now. This equation is homogeneous, so try substituting equation (5) into equation (2).

$$LAp^2 e^{pt} + RAp e^{pt} + \frac{1}{C} A e^{pt} = 0. \tag{7}$$

We want a solution for all times t, but this can be true since Ae^{pt} can be canceled out, providing that the remaining equation,

$$Lp^2 + Rp + \frac{1}{C} = 0, \tag{8}$$

can be satisfied. This is a quadratic equation and can be solved immediately to give

$$p_{1,2} = \frac{-R \pm \sqrt{R^2 - (4L/C)}}{2L}. \tag{9}$$

Apparently we have two solutions, but this is not surprising, since a second-order differential equation would need to be integrated twice in principle, and each time would involve one arbitrary constant of integration which would multiply each of the two solutions, like this

$$Q_1(t) = A_1 e^{p_1 t}, \quad Q_2(t) = A_2 e^{p_2 t}. \tag{10}$$

The general solution is a sum of the two:

$$Q(t) = \exp\left(-\frac{R}{2L}t\right) \left\{ A_1 \exp\left[\sqrt{\left(\frac{R}{2L}\right)^2 - \frac{1}{LC}}\, t\right] \right.$$
$$\left. + A_2 \exp\left[-\sqrt{\left(\frac{R}{2L}\right)^2 - \frac{1}{LC}}\, t\right] \right\}. \tag{11}$$

Like all good manipulators of equations, we should simplify our notation, and we could set $R/2L = \gamma$ and set $1/LC = \omega_0^2$. We could even call $\gamma^2 - \omega_0^2 = (\omega')^2$. Then (11) becomes

$$Q(t) = e^{-\gamma t} \left(A_1 e^{\sqrt{(\omega')^2}\, t} + A_2 e^{-\sqrt{(\omega')^2}\, t} \right). \tag{12}$$

Now we must have a mathematical digression. The reason is that $(\omega')^2$ may be either positive or negative, depending on whether γ is greater than or

Learning Guide 18   AC Circuits

less than ω_0. In the former case there is no problem, because γ is greater than ω' and we can see that the solution will just be a damped exponential with a damping function of either

$$Q_1(t) \propto e^{(-\gamma+\omega')t}$$

or (13)

$$Q_2(t) \propto e^{(-\gamma-\omega')t}.$$

These solutions are called **overdamped** and would occur when R is large in the circuit.

Another possibility is that ω_0 is exactly equal to γ. In that case the two solutions Q_1 and Q_2 are identical, and the system damps down as follows:

$$Q_1(t) = Q_2(t) \propto e^{-\gamma t}. \tag{14}$$

This is called **critical damping**.

But what about the most common case in electrical circuits, where the resistance is not dominant and hence ω_0 is larger than $\gamma = R/2L$? In this case, (ω'^2) is negative, and so we must write $(\omega')^2 = i|\omega'|$, where i (or j, as most engineers write it) is $\sqrt{-1}$ and where

$$|\omega'| = \sqrt{\omega_0^2 - \gamma^2}. \tag{15}$$

Thus, our solution in this case, which we can call **underdamped**, or oscillatory, can be written as

$$Q(t) = e^{-\gamma t}(A_1 e^{i\omega' t} + A_2 e^{-i\omega' t}). \tag{16}$$

To make this solution more physical and comprehensible, we want to take advantage of the *great* and perhaps *most remarkable equation* in mathematics

$$e^{i\theta} = \cos\theta + i\sin\theta, \tag{17}$$

or, if we want to make it a function of time,

$$e^{i\omega t} = \cos(\omega t) + i\sin(\omega t). \tag{18}$$

This strange equation (which has one exotic result that the combination of two transcendental numbers and an imaginary number gives a simple integer, that is, $e^{i\pi} = -1$) is true, because it can easily be shown that

$$e^\theta = 1 + \theta + \frac{\theta^2}{2!} + \frac{\theta^3}{3!} + \frac{\theta^5}{5!} + \cdots \tag{19}$$

and

$$\cos\theta = 1 - \frac{\theta^2}{2!} + \frac{\theta^4}{4!} - \cdots \tag{20}$$

and

$$\sin\theta = \theta - \frac{\theta^3}{3!} + \frac{\theta^5}{5!} - \cdots \tag{21}$$

by examining the Maclaurin series for these functions and comparing the even and odd terms of (19) with (20) and (21).

This relation is exactly what is needed to give an engineer or a physicist a useful graphical picture of the amplitude and phase of oscillatory behavior. The cosine and sine that appear in (18) immediately evoke the accompanying figure.

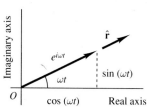

This picture represents a rotating vector along the direction marked $\hat{r}$. The real function cosine is like a projection along the real axis, while the sine function represents a vertical component along the imaginary direction. In other words, the use of i in the mathematics almost automatically creates a sort of two-dimensional vector space that replaces the necessity for using unit vectors like $\hat{x}$ and $\hat{y}$.

Now, we can think like realists, physicists, and engineers and know that what we want to get out of this mathematics are real numbers that represent real things like current and voltage. So we can define the real (Re) part of this complex "vector" and the imaginary (Im) part as follows:

$$\text{Re}\left[e^{i\omega t}\right] = \cos(\omega t) \tag{22}$$

and

$$\text{Im}\left[e^{i\omega t}\right] = \sin(\omega t). \tag{23}$$

Learning Guide 18 AC Circuits

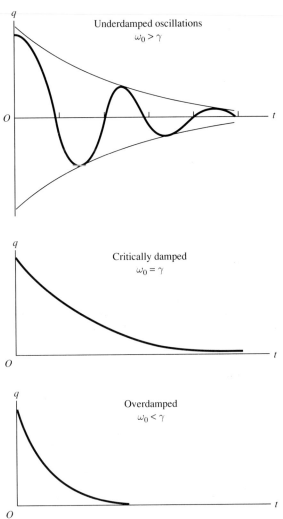

And we can also represent trigonometric functions with complex numbers by inverting equation (18):

$$\cos(\omega t) = \frac{e^{i\omega t} + e^{-i\omega t}}{2}; \tag{24}$$

$$\sin(\omega t) = \frac{e^{i\omega t} - e^{-i\omega t}}{2i}. \tag{25}$$

At this point we can make simple sense out of equation (16). The easiest (and trickiest) way to simplify is to call

$$A_1 = \frac{1}{2}A_0 e^{i\theta} \quad \text{and} \quad A_2 = \frac{1}{2}A_0 e^{-i\theta}. \tag{26}$$

Then, using (24), we immediately have

$$Q_{\text{physical}}(t) = \text{Re}\,[Q(t)] = A_0 e^{-\gamma t}\cos(\omega' t + \theta), \tag{27}$$

in which A_0 plays the familiar role of the amplitude and θ is the phase, both of which can be determined from the so-called initial conditions.

As an example, suppose we want to know what the voltage is across the capacitor C in our example at any time. And suppose that at $t = 0$ we start with no current flowing anywhere, but with the capacitor charged to voltage V_0. Then the voltage across the capacitor, written in terms of (27), is

$$V(t) = \frac{Q(t)}{C} = \frac{A_0}{C} e^{-\gamma t} \cos(\omega' t + \theta).$$

At $t = 0$ this gives

$$V_0 = \frac{Q(0)}{C} = \frac{A_0}{C} \cos\theta. \tag{28}$$

We must have also have no initial current, so that $I(0) = dQ/dt = 0$. Since

$$I(t) = \frac{dQ}{dt} = -A_0[\gamma e^{-\gamma t}\cos(\omega' t + \theta) + e^{-\gamma t}\omega'\sin(\omega' t + \theta)];$$

at $t = 0$ this gives

$$0 = \gamma \cos\theta + \omega' \sin\theta. \tag{29}$$

From (28) and (29) we can find the phase and amplitude, but it is not too simple. However, when γ is small, we can see by inspection that $\theta = 0$, and $A_0 \simeq CV_0$, in direct analogy to a mass on a spring that is held displaced and then let go at $t = 0$.

Complex Impedance

Now we want to turn to another very interesting situation in which we have an **inhomogeneous** equation. Very often in physics and engineering we are interested in the steady-state electrodynamics of our RCL circuit when it is plugged into an AC generator as shown in the figure.

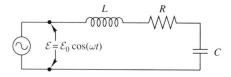

In other words, the equation of the RCL circuit is no longer homogeneous, because there is a voltage impressed on the terminals:

$$L\frac{d^2 Q}{dt^2} + R\frac{dQ}{dt} + \frac{1}{C}Q = \mathcal{E}_0 \cos(\omega t). \tag{30}$$

This is an inhomogeneous equation in which the term K in equation (4) is a function of time. That this can be solved is an extension of our previously mentioned theorem and can be stated this way: the solution of an *inhomogeneous equation* can be found to be the *general solution* of the homogeneous equation

Learning Guide 18   AC Circuits

added to a special, or *particular, solution* of the inhomogeneous equation. That is, we can write the solution similar to (6) as

$$Q(t) = Ae^{pt} + B(t), \tag{31}$$

where we have written the special solution of the inhomogeneous equation as $B(t)$—that is, as a function of time. Since we already have a general solution to equation (2), let's try to find the special solution $B(t)$ of equation (31).

A first valuable trick is to take a hint from the previous work and decide that it might be mathematically simpler to represent the driving voltage on the circuit, not as the real function $\mathcal{E}_0 \cos(\omega t)$, but as the complex function $\mathcal{E}_0 e^{i\omega t}$, keeping in mind that we will ultimately want to take a real part of this solution to represent reality. Thus we want to solve the equation

$$L\frac{d^2Q}{dt^2} + R\frac{dQ}{dt} + \frac{1}{C}Q = \mathcal{E}_0 e^{i\omega t}. \tag{32}$$

Because we know that exponential functions have derivatives that still contain the exponential function, why not try

$$Q(t) = B(t) = B_0 e^{i\omega t}? \tag{33}$$

Thus, we substitute it into equation (32) and find that

$$-LB_0 \omega^2 e^{i\omega t} + iR\omega B_0 e^{i\omega t} + \frac{B_0}{C} e^{i\omega t} = \mathcal{E}_0 e^{i\omega t}. \tag{34}$$

This equation clearly holds true all the time, too, because we can divide out the factor $e^{i\omega t}$ and satisfy the equation by satisfying

$$-LB_0 \omega^2 + iR\omega B_0 + \frac{B_0}{C} = \mathcal{E}_0. \tag{35}$$

In fact, we can find B_0 to be just

$$B_0 = \frac{\mathcal{E}_0}{-L\omega^2 + iR\omega + 1/C} \tag{36}$$

by rearrangement.

However, we are not usually very interested in Q but rather in $I = dQ/dt$. So we differentiate equation (33) and find the current to be

$$I(t) = \frac{dQ}{dt} = i\omega B_0 e^{i\omega t} \equiv I_0 e^{i\omega t}. \tag{37}$$

By dividing this equation into (34), we can get an equation for $\mathcal{E}_0/I_0$:

$$iL\omega + R - \frac{i}{\omega C} = \frac{\mathcal{E}_0}{I_0}. \tag{38}$$

At this point we are in a position to make a *grand generality*. Think of Ohm's law, which predicts a passive response to electromotive force through a resistor as being a flow of current that is linearly proportional to the applied electromotive force. In fact, equation (33) would reduce to Ohm's law if we had no inductance L and if we shorted out the capacitance C—i.e., we would get

$$R = \frac{\mathcal{E}_0}{I_0}. \tag{39}$$

In general, for AC (alternating-current) circuits, we think of the terms on the left side of equation (38) as forming a complex number which consists of a real **resistance** R and imaginary terms called **reactances** defined by $X = \omega L - (1/\omega C)$. These go together to make up an **impedance** Z, which can be represented by the simple linear relation

$$Z = \frac{\mathcal{E}_0}{I_0} = R + iX. \tag{40}$$

Since Z is complex, so, in general, are $\mathcal{E}_0$ and I_0.

We can then work out values for Z, $\mathcal{E}$, and I, assuming them to be complex numbers. At the end, we can simply take the real part of I_0 and $\mathcal{E}_0$ as the actual physical quantity we observe. When we do this, we obtain a cosine function that contains the correct amplitude and phase information.

From now on, when we are working with a quantity as a complex number we will put a small, wiggly line (a tilde) over it so that we know that it is complex. At the end we will have to take the real part to get our answer for I or $\mathcal{E}$.

To make these ideas clear, let's consider a number of simple circuits and imagine applying a voltage $\mathcal{E}_0 e^{i\omega t}$ to each of them. However, let's draw a snapshot picture of what's going on in the complex plane at the precise instant when the voltage vector, which is thought to be rotating, has been caught at the real axis (or x-axis).

Example 1:

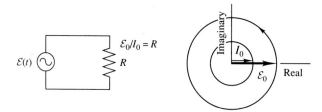

We see that the current and the voltage are in phase with one another and that

$$I(t) = \mathrm{Re}\left[\frac{\mathcal{E}_0 e^{i\omega t}}{R}\right] = \frac{\mathcal{E}_0}{R}\cos(\omega t). \tag{41}$$

Learning Guide 18 AC Circuits

Example 2:

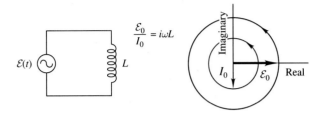

$$I(t) = \text{Re}\left[\frac{\mathcal{E}_0 e^{i\omega t}}{i\omega L}\right] = \text{Re}\left[-i\frac{\mathcal{E}_0}{\omega L}e^{i\omega t}\right] = \frac{\mathcal{E}_0}{\omega L}\sin(\omega t). \qquad (42)$$

We see that the current lags in phase 90° behind emf, and this is exactly what one expects the "inertia" of the inductance to do. That is, just because you apply a "force" to a "mass" does not mean that its "velocity" will immediately be different from zero.

Example 3:

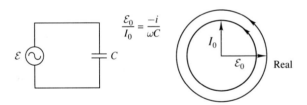

$$I(t) = \text{Re}[i\mathcal{E}_0 \omega C e^{i\omega t}] = -\mathcal{E}_0 \omega C \sin(\omega t). \qquad (43)$$

That is, the current leads by 90°, which is to say that the current must flow into a capacitor first before it can charge up a voltage different from zero.

We can now state general principles for solving problems with complex numbers:

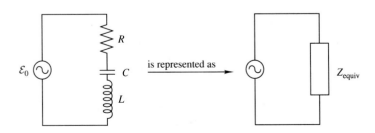

Work out Z from:

$$\text{Series:} \quad Z_{\text{equiv}} = Z_1 + Z_2 + \cdots,$$

$$\text{Parallel:} \quad \frac{1}{Z_{\text{equiv}}} = \frac{1}{Z_1} + \frac{1}{Z_2} + \cdots,$$

where Z is a complex number made up from

$$Z_R = R \quad \text{(real)}, \qquad Z_C = \frac{1}{i\omega C} = \frac{-i}{\omega C} \quad \text{(imaginary), and}$$

$$Z_L = i\omega L \quad \text{(imaginary)}.$$

A complex number

$$\widetilde{Z} = R + iX$$

may be written as

$$\widetilde{Z} = |Z|e^{i\theta},$$

where

$$|Z| = \sqrt{R^2 + X^2}$$

and

$$\theta = \tan^{-1}\left(\frac{X}{R}\right).$$

With this approach, we can work out Z for a complicated circuit by using complex numbers and then, at the end, find the magnitude and phase as follows:

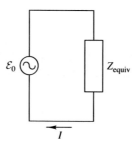

As an example of how to do this, let's consider a series RLC circuit.

Learning Guide 18 AC Circuits

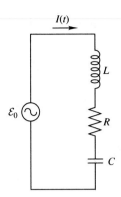

Here, $\widetilde{Z} = R - (i/\omega C) + i\omega L = R + i\,[\omega L - (1/\omega C)]$, which we can write as

$$\widetilde{Z} = |Z|e^{i\theta},$$

where

$$|Z| = \sqrt{R^2 + \left(\omega L - \frac{1}{\omega C}\right)^2}$$

and

$$\theta = \tan^{-1}\left[\frac{\omega L - (1/\omega C)}{R}\right].$$

If we express

$$I(t) = \operatorname{Re}\left[\frac{\widetilde{\mathcal{E}}}{\widetilde{Z}}\right] = \operatorname{Re}\left[\frac{\mathcal{E}_0 e^{i\omega t}}{|Z|e^{i\theta}}\right] = \operatorname{Re}\left[\frac{\mathcal{E}_0 e^{i(\omega t - \theta)}}{|Z|}\right],$$

then

$$I(t) = \frac{\mathcal{E}_0}{|Z|}\cos(\omega t - \theta)$$

$$= \frac{\mathcal{E}_0}{|Z|}[\sin(\omega t)\sin\theta + \cos(\omega t)\cos\theta]$$

$$= \mathcal{E}_0\left\{\frac{[\omega L - (1/\omega C)]\sin(\omega t) + R\cos(\omega t)}{R^2 + [\omega L - (1/\omega C)]^2}\right\}.$$

In this way we can derive the amplitude and phase relationships for current and voltage in any arbitrary combinations of resistors, capacitors, and inductors. We can also see easily the phase relations between the voltages and currents in the individual components of our circuit.

$$\widetilde{\mathcal{E}} = \widetilde{\mathcal{E}}_R + \widetilde{\mathcal{E}}_C + \widetilde{\mathcal{E}}_L$$

$$= \widetilde{Z}\widetilde{I}$$

$$= [R - (i/\omega C) + i\omega L]\,\widetilde{I}.$$

Noting that

$$e^{i\pi/2} = \cos\left(\frac{\pi}{2}\right) + i\sin\left(\frac{\pi}{2}\right) = i$$

and

$$e^{-i\pi/2} = \cos\left(\frac{\pi}{2}\right) - i\sin\left(\frac{\pi}{2}\right) = -i,$$

we may then write

$$\widetilde{\mathcal{E}} = R\widetilde{I} + \frac{e^{-i\pi/2}}{\omega C}\widetilde{I} + e^{i\pi/2}(\omega L)\widetilde{I}.$$

Note: In the first term on the right-hand side, $\widetilde{\mathcal{E}}_R = R\widetilde{I}$ is in phase with $\widetilde{I}$; in the second term, $\widetilde{\mathcal{E}}_C$ lags $\widetilde{I}$ by 90°; in the last term, $\widetilde{\mathcal{E}}_L$ leads $\widetilde{I}$ by 90°.

These phase relationships make sense if you think about the physics of the devices. The current flows through a resistor the instant you apply a voltage. However, with a capacitor, you must flow current for a while to build up an appreciable voltage, so the voltage lags behind the current. With an inductor, if you suddenly apply an external voltage, the current takes a while to build up, because the induced magnetic field opposes the buildup according to Lenz's law. So, in an inductor the voltage leads the current.

Resonance

Now let's look again at the *RCL* circuit. We can write equation (38) directly from our idea that each circuit element has a complex impedance:

$$I_0 = \frac{\mathcal{E}_0}{Z} = \frac{\mathcal{E}_0}{R + i\,[\omega L - (1/\omega C)]} \tag{44}$$

or

$$I(t) = \text{Re}\left\{\frac{\mathcal{E}_0}{R + i\,[\omega L - (1/\omega C)]} e^{i\omega t}\right\}. \tag{45}$$

We can see by inspection of this equation that, at a very low frequency, $1/\omega C$ becomes very large and the current drops to the low value as given by (43) and leads the voltage. At very high frequencies, the ωL becomes largest, and the current has the low value as given by (42) and lags behind the voltage.

Learning Guide 18   AC Circuits

At **resonance** the current becomes a maximum, because the reactances of capacitor and inductor cancel each other and we have

$$I_0 = \frac{\mathcal{E}_0}{R}$$

when

$$\omega L = \frac{1}{\omega C}, \quad \text{or} \quad \omega = \omega_0 = \frac{1}{\sqrt{LC}}. \tag{46}$$

If we want to do the sort of algebra that enables us to write $I(t)$ explicitly as a real and imaginary part, we can multiply both the top and the bottom of the right-hand side of equation (44) by $R - i[\omega L - (1/\omega C)]$ to get

$$I_0 = \left\{ \frac{R}{R^2 + [\omega L - (1/\omega C)]^2} - \frac{i[\omega L - (1/\omega C)]}{R^2 + [\omega L - (1/\omega C)]^2} \right\} \mathcal{E}_0. \tag{47}$$

We can describe the first part as the "in-phase" component of the current response to the force, and the second part as the "out-of-phase" component. Every resonance in nature has this property, whether it is a mass on a spring, the gyration of a nucleus in a magnetic resonance experiment, a dust particle in a light beam, a radio antenna, or the response of a strange elementary particle to a high-energy gamma ray.

To illustrate how we should write (47) in completely real form, we must say that

$$I(t) = \text{Re}\left[\left\{ \frac{R}{R^2 + [\omega L - (1/\omega C)]^2} - \frac{i[\omega L - (1/\omega C)]}{R^2 + [\omega L - (1/\omega C)]^2} \right\} \mathcal{E}_0 e^{i\omega t}\right] \tag{48}$$

If we want the real physical version of this equation, we use (18), to get

$$I(t) = \frac{R\cos(\omega t) + [\omega L - (1/\omega C)]\sin(\omega t)}{R^2 + [\omega L - (1/\omega C)]^2} \mathcal{E}_0. \tag{49}$$

This is the same result as derived previously by using Z and θ.

It is easier in general, however, not to write out this equation in full, but to use our physical imagination on the rotating imaginary-plane picture that follows, remembering that we have made a snapshot of the current and voltage at a particular instant and that the whole picture can be thought of as rotating around O in a counterclockwise direction with angular velocity ω. We must imagine projecting the complex vectors down onto the real axis.

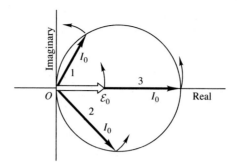

Complex-plane "snapshot" of I_0 and $\mathcal{E}_0$

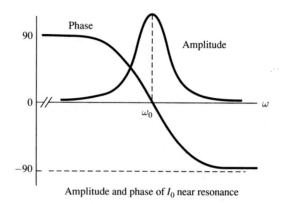

Amplitude and phase of I_0 near resonance

Note that the amplitude is

$$|I| = \frac{\mathcal{E}_0}{|Z|} = \frac{\omega \mathcal{E}_0}{L\sqrt{(\omega^2 R^2/L^2) + (\omega^2 - \omega_0^2)^2}}$$

and that the phase is given by

$$\phi = -\tan^{-1}\left[\frac{L(\omega_0^2 - \omega^2)}{\omega R}\right],$$

where $\omega_0 = 1/\sqrt{LC}$.

Power and Quality Factor

We know that current times voltage gives power in watts. However, suppose that we want to know the power dissipation in a reactive or imaginary part of a circuit. Let there be a current I flowing through the circuit element. According to our generalization, the *voltage drop* across that element will be IZ. If Z is determined by a reactive element, then $Z = iX$, and V will be equal to iIX. The voltage will be out of phase with the current. But we must think of actual real quantities, so the power must be $\text{Re}[I] \times \text{Re}[V]$. What this represents is a situation where the current has a real component that is proportional to, say, $\cos(\omega t)$, but where the voltage will be represented by a real component that is out of phase, namely, by $\sin(\omega t)$. But on the average,

$$\int_0^\infty \sin(\omega t)\cos(\omega t)\,dt$$

is always zero, leading to the result that there is no average power dissipated in reactive elements. Of course, energy does indeed flow into and out of such a circuit element during each cycle.

On the other hand, since current and voltage are in phase in resistive or real elements, the power average will be an average of $\cos^2(\omega t)$ or $\sin^2(\omega t)$, which you know to be $\frac{1}{2}$. So the power dissipated in a resistor is given by

$$P = \tfrac{1}{2}|I_0|^2 R = \tfrac{1}{2}|I|^2 R = \tfrac{1}{2}RI^*I, \tag{50}$$

where the asterisk means the complex conjugate of the quantity, obtained by reversing the sign of the imaginary part, and symbolized by reflecting the imaginary pictures we have drawn in the x-axis.

We can write the power in these circuits in other ways. For example,

$$RI^*I = I^*I\,\text{Re}[Z],$$

since I^*I is already real and $R = \text{Re}[Z]$ and thus $I^*I\,\text{Re}[Z] = \text{Re}[I^*IZ] = \text{Re}[I^*V]$, leading to

$$P = \tfrac{1}{2}\text{Re}[I^*V] \quad \text{or} \quad \tfrac{1}{2}\text{Re}[IV^*]. \tag{51}$$

Using the foregoing equations you can also show that

$$P = \frac{1}{2}V^*V\,\text{Re}\left[\frac{1}{Z}\right]. \tag{52}$$

Because of the factor of $\frac{1}{2}$ that is present in all of these equations, one frequently absorbs that factor into the current or voltage by defining an I_{rms} or V_{rms} which is just $1/\sqrt{2}$ times less than the I_0 or V_0 we have used to represent the amplitude or maximum value of our complex vectors. Therefore, if you were to look at 120 volts from a household circuit on your laboratory oscilloscope, you would see that it would actually reach a peak level of 169.7 volts at the maximum of the sine curve on the scope.

The **quality factor** of a resonant circuit is called Q and can be defined as the ratio of the resonance frequency to the difference between the two frequencies above and below resonance at which the power dissipated in the resonant circuit is $\frac{1}{2}$ as much as that exactly at resonance:

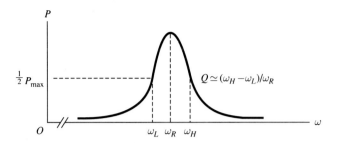

It must remain as a project for those interested to show that

$$Q = \frac{\omega L}{R} = \frac{\omega_R}{\omega_H - \omega_L} \tag{53}$$

and, using the middle expression above, to show that in the case of a damped oscillator it will oscillate for Q cycles before its amplitude has diminished to less than about 5% of the original. The Q for a coil or cavity that is slightly lossy can also be defined as the ratio of the average energy stored in the cavity to the energy dissipated in a cycle, but this involves calculations that are a little tedious to carry out. At any rate, when one talks about a high-Q circuit, one means a very sharply tuned device of narrow frequency response.

learning guide 19

Maxwell's Equations and Electromagnetic Waves

Suggested Reading: Fishbane, Gasiorowicz, & Thornton, Chapter 35, and the supplemental notes that follow this Learning Guide

PROBLEM I

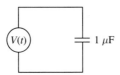

1. Show that the displacement current in a parallel-plate capacitor can be written as

$$I_d = C \frac{dV}{dt}.$$

   (You should be able to do this without help. If not, you may look at Helping Questions 1 and 2.)

2. What voltage $V(t)$ would you use to get the displacement current shown in a 1-μF capacitor? See Helping Question 3 if you need assistance.
   Key 12

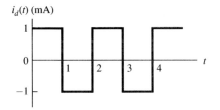

PROBLEM II

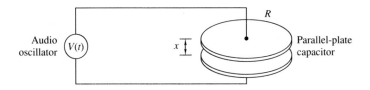

A lecture demonstration of the displacement current in a parallel-plate capacitor is sketched in the figure. The demonstration was first done 60 years after Maxwell's prediction. The apparatus shown is one built by Professor Thomas Carver and Jan Rajhel of Princeton University.*

$$V(t) = V_m \sin(\omega t), \quad V_m = 100 \text{ V}, \quad \omega = 1.2 \times 10^5 \text{ rad/s},$$

$$R = 40 \text{ cm}, \quad x = 10 \text{ cm}$$

1. Ignore fringing fields and calculate the displacement current between the plates. (Be sure to use the correct units.) **Key 25**
2. Find the **B**-field at a distance $r = 38$ cm from the centerline of the plates. (You may need Helping Questions 4 and 5.) **Key 10**

The hard part of this experiment is to detect such a weak displacement current, spread out over such a large area. (That's what took 60 years in the early days of electronics.) Carver did it by winding a remarkable coil to detect the **B**-field associated with I_d. As shown in the next figure, the coil is wound on

---

* For additional details, see *American Journal of Physics*, Vol. 42, p. 246 (1974).

a toroidal form, which is then slid between the capacitor plates. An oscillating voltage $V(t)$ appears on the oscilloscope.

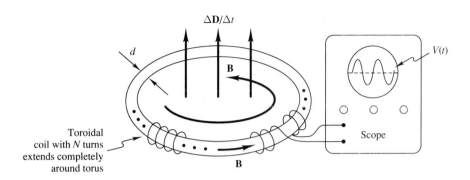

3. Which of Maxwell's equations is exploited to measure B and subsequently I_d? **Key 17**
4. What is the voltage $V(t)$ read on the scope? (You probably don't need Helping Question 6.) **Key 24**
5. If the torus is removed from between the capacitor plates and instead encircles the wire between the oscillator and the capacitor, what is the new $V(t)$? **Key 6**

PROBLEM III

1. For a plane electromagnetic wave traveling in the $+x$-direction (i.e., the **i** direction), what is the direction of **B** when $\mathbf{E} = E\mathbf{j}$? When $\mathbf{E} = E\mathbf{k}$? **Key 14**

   (For help, look at Fishbane, Gasiorowicz, & Thornton, Figures 35-6 and 35-7.)

2. Find $|\mathbf{S}(x,t)|$ for the standard wave functions (like Fishbane, Gasiorowicz, & Thornton, Eq. (35-8) and Eq. (35-13), except in the x direction) in terms of E_0, x, and t. What is the time-averaged flux $\langle S \rangle$ passing a given point, say, $x = 0$? **Key 5**

   (Consult Helping Question 7 for the second part.)

3. A New York City FM station is 100 km from Princeton, New Jersey, and radiates an *average* power of 50 kW. Assuming spherical radiation, what is the value of S in Princeton? What is the magnitude of E_0 at an antenna that is located there? (Be sure the units work!) Finally, what is B_0? **Key 16**

4. FM radio frequencies are about 100 MHz. What is the wavelength? An electric dipole antenna (see Fishbane, Gasiorowicz, & Thornton, Figure 35-14, and remember that an antenna that works as a transmitter will also work as a receiver) has a length of $\lambda/2$. Estimate the signal voltage across

such an antenna from your answer to part (3). Look at the noise voltage specification on an FM tuner. Is there enough signal to overcome the noise?
Key 23

PROBLEM IV

The difference in wavelength between an incident microwave beam and the beam reflected from an approaching or receding car is used to determine automobile speeds on the highway.

1. Show that if v is the speed of the car and f the frequency of the incident beam, the change of frequency is approximately $2fv/c$, where c is the speed of the electromagnetic radiation.

   (If you have difficulty, see Helping Questions 8 through 10.)

2. For microwaves of frequency 2450×10^6 Hz, what is the change in frequency per mi/h of speed? **Key 21**

PROBLEM V

In this problem we will investigate the conditions on a small speck of cosmic dust that will determine whether it will or will not be blown away by the sun's radiation pressure. Assume for simplicity that the dust particle moves only in a radial direction with respect to the center of the sun and that it is at a distance r from the sun.

1. What is the gravitational force on the particle? **Key 7**
2. If the radiant power emitted by the sun is P_0, what is the radiant intensity a distance r from the sun? (See Helping Question 11 for assistance.) **Key 35**
3. Assuming the dust particle is a sphere of radius a and that it absorbs all radiant energy incident upon it, how much energy per unit time does it absorb? **Key 22**

   (You probably don't need Helping Questions 12 and 13.)

4. What force is exerted by the light that the dust particle absorbs? **Key 40**

   (Stuck? Try Helping Questions 14, 15, and 16, one at a time.)

5. What is the *net* radial force acting on the dust particle? **Key 49**
6. What conditions must a and m satisfy to ensure that the particle will be blown *away* from the sun? **Key 4**
7. Will a small sphere of ice weighing 1 μg be blown out of the Solar System? **Key 44**

Numerical data:

$$c = 3 \times 10^8 \text{ m/s},$$
$$P_0 = 3.9 \times 10^{26} \text{ W}$$
$$M_\odot = 2 \times 10^{30} \text{ kg}$$
$$\rho_{\text{ice}} = 1 \text{ g/cm}^3 = 10^3 \text{ kg/m}^3$$
$$G = 6.67 \times 10^{-11} \text{ N} \cdot \text{m}^2/\text{kg}^2$$

PROBLEM VI

A cylindrical resistor has length l, radius a, and resistivity ρ; it carries a constant current I. Its leads are cylindrical wires of radius r_0 and are perfect conductors, that is, have zero resistivity. The voltage drop across the resistor is V.

1. Find the Poynting vector **S** (magnitude and direction) at the surface of the resistor. **Key 41**

   If you have trouble, review Fishbane, Gasiorowicz, & Thornton, Section 35-4 and see Helping Question 17.

2. Show that the rate P at which energy flows into the resistor through its cylindrical surface is equal to the rate at which Joule heat is produced.

   See Helping Questions 18 and 19 for aid.

3. Find the Poynting vector at the surface of the wire. What is the energy flow in the wire? **Key 19**

   See Helping Question 20 if you don't understand.

4. What do the results of parts (3) and (2) imply? **Key 46**

PROBLEM VII

1. Two ideal polarizers (1 and 3) are set up with their polarization directions at right angles to each other. Another (2) is located between the first two, with its polarization at an angle ϕ to that of the first. If the incident light (before the first polarizer) is unpolarized and has intensity I_0, what is the intensity I_1 after the first polarizer? **Key 28**

   If you're having difficulty, try Helping Questions 21 through 23.

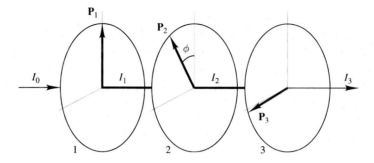

2. What is the intensity I_2 after the second polarizer? **Key 45**

You will probably need Helping Questions 24 and 25.

3. What is the intensity I_3 after the light wave passes through all three polarizers? **Key 42**

Use Helping Question 26 reluctantly.

4. What happens to the intensity when the middle polarizer is removed? **Key 36**

HELPING QUESTIONS

1. What is the capacitance of a parallel-plate capacitor in terms of its area A and plate separation x? **Key 15**

2. What is E in terms of V and x, and Φ_E in terms of E and A? Now, put it all together. **Key 2**

3. From your result in part (1), find the $V(t)$ required to give $I_d = +1$ mA. **Key 26**

4. Symmetry says that B_ϕ is constant on circles around the centerline. Evaluate $\oint \mathbf{B} \cdot d\mathbf{l}$ on a circle of radius r. **Key 20**

5. What is the displacement current flowing through a circle of radius r? **Key 32**

6. How is the voltage across the coil related to B_ϕ? **Key 13**

7. What is the value of the integral of $\sin^2 \theta$ over one cycle? Alternatively, plot $\sin^2 \theta$ versus θ and read off the average value. **Key 8**

8. Assuming that the car is coming toward the speed trap, what frequency f_c would a microwave detector mounted on the moving car receive? **Key 1**

9. What is the frequency f_r reflected from the moving car in the police car's reference frame? **Key 18**

10. What is the approximate frequency f_d detected by the detector? **Key 31**

11. Assume P_0 is uniformly distributed over a spherical surface of radius r.

12. What is the cross-sectional area of the dust sphere? **Key 3**

13. How much energy per unit time passes through area πa^2 in terms of the intensity $\langle S(r) \rangle$? **Key 27**
14. How much momentum is transferred to the particle when an amount U of radiant energy is absorbed? **Key 11**
15. What is the rate of change of the momentum associated with light absorption? **Key 37**
16. What is the radiant force on the particle? **Key 29**
17. What are the electric and the magnetic induction fields at the surface of the resistor? **Key 33**
18. What is the energy flux through the cylindrical surface of the resistor per unit time? **Key 9**
19. What is the rate of Joule heating? **Key 48**
20. What are the E- and B-fields at the surface of the wire? Why? **Key 43**
21. What does "unpolarized" mean? If you are not sure, see Fishbane, Gasiorowicz, & Thornton, Figure 35-19, for a picture. Now try the problem again.
22. Still uncertain? Is the intensity of the incoming light with $\mathbf{E}_\perp$ along $\mathbf{P}_1$ different from the intensity of the light with $\mathbf{E}_\perp$ perpendicular to $\mathbf{P}_1$? **Key 38**
23. What does polarizer 1 do to the $\mathbf{E}_\parallel$ part of the wave? To the $\mathbf{E}_\perp$ part? Now reason out the answer. **Key 30**
24. Basically, does the polarizer work on the E-field or on the intensity? Go on to Helping Question 25. **Key 47**
25. Take the E-field coming into polarizer 2 and resolve it parallel to $\mathbf{P}_2$ and perpendicular to $\mathbf{P}_2$. What is the E-field that gets through? What intensity gets through? **Key 34**
26. What is the relationship between $|\mathbf{E}_2|$ and $|\mathbf{E}_3|$? I_2 and I_3? **Key 39**

Notes: How to Get Light Waves from Maxwell's Equations

This reading will supplement, with a slightly different point of view, the discussion of electromagnetic waves in Fishbane, Gasiorowicz, & Thornton. It represents as deep and fundamental a penetration into the mathematical meaning of Maxwell's equations and electromagnetic waves as we are able to make in these Learning Guides.

Two of Maxwell's equations are simple and not important for this topic, although they are important for electromagnetostatics:

Gauss' law:

$$\epsilon_0 \oiint \mathbf{E} \cdot d\mathbf{A} = \iiint_V \rho \, dV = Q_{\text{inside}}; \tag{1}$$

$$\oiint \mathbf{B} \cdot d\mathbf{A} = 0.* \tag{2}$$

---

* The double and triple integral signs denote surface and volume integrals, respectively. The symbols $\oint$ and $\oiint$ indicate integration over closed paths or surfaces.

Equation (1) shows how the $\vec{\mathbf{E}}$-field is produced by charges; (2) indicates the nonexistence of analogous magnetic charges.

The following two of Maxwell's equations will be needed in what follows:

Ampère's law with displacement current:

$$\oint \mathbf{B}\cdot d\mathbf{s} = \mu_0 \left[I + \epsilon_0 \frac{d\Phi_E}{dt} \right] \qquad (3)$$

Faraday's law:

$$\oint \mathbf{E}\cdot d\mathbf{s} = -\frac{d\Phi_B}{dt} \qquad (4)$$

The meaning of the terms on the right-hand side is as follows:

$$\Phi_E = \iint \mathbf{E}\cdot d\mathbf{A} \qquad (5)$$

$$\Phi_B = \iint \mathbf{B}\cdot d\mathbf{A} \qquad (6)$$

$$I = \iint \mathbf{J}\cdot d\mathbf{A}. \qquad (7)$$

In other words, these terms represent the electric field flux, the magnetic field flux, and the current density through the surfaces d**A**, which are bounded by the paths d**s** along which **E** and **B** are integrated in the operations represented by the symbols on the left-hand side of equations (3) and (4).

When we have a nonconducting medium, no current flows, and we can write the two Maxwell equations in a nearly symmetrical form. These two equations represent the starting point of our description of waves:

$$\oint \mathbf{E}\cdot d\mathbf{s} = -\iint \frac{\partial \mathbf{B}}{\partial t}\cdot d\mathbf{A} \qquad (8)$$

$$\oint \mathbf{B}\cdot d\mathbf{s} = \mu_0 \epsilon_0 \iint \frac{\partial \mathbf{E}}{\partial t}\cdot d\mathbf{A} \qquad (9)$$

You should notice that we have taken partial derivatives of **E** and **B** instead of taking the derivative of the entire flux integral. This is because in what follows we are assuming that the path and the surface through which we have a changing flux do not change with time; that is, only the electromagnetic fields **E** and **B** change with time.

Assume first of all that we have chosen the coordinate system shown in diagram 1 in which there is an electric field **E** in the y-direction that is not a constant as far as x is concerned and also depends on time. We also assume there is a **B**-field in the z-direction. That is,

$$\mathbf{E} = E_y(x,t)\,\mathbf{j}; \qquad \mathbf{B} = B_z(x,t)\,\mathbf{k}.$$

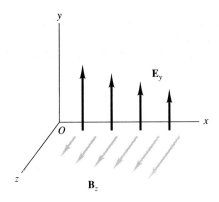

Diagram 1

Now consider a small path as shown by diagram 2, which we use to take a circuital path of integration as specified by the left-hand side of (8).

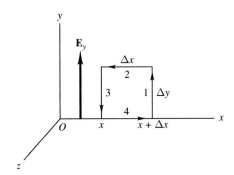

Diagram 2

The left-hand side of Maxwell's equation (8) is evaluated as follows:

$$\oint \mathbf{E} \cdot d\mathbf{s} = E_y(x + dx)\Delta y + 0 - E_y(x)\Delta y + 0. \tag{10}$$

Note that, on the right-hand side, the first term is the contribution to the line integral from side 1 of the square, the second term from side 2, and so on. The right-hand side of equation (8) can be evaluated for the same rectangle as indicated by diagram 3.

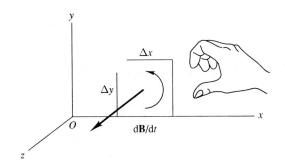

Diagram 3

Notice that (8) uses the fact that there is also a **B**-field that has a z-component, and, moreover, that the rate of change of this **B**-field is positive in the $+z$-direction. Notice also that we have used a right-hand rule in diagram 2 in order to relate the sense of direction as to how we should go around the path of integration when the rate of change of **B** is in the $+z$-direction, as shown in diagram 3. The result is

$$-\iint \frac{\partial \mathbf{B}}{\partial t} \cdot d\mathbf{A} \simeq -\left.\frac{\partial B_z}{\partial t}\right|_{x+(\Delta x/2)} \Delta x \Delta y. \qquad (11)$$

Combining (10) and (11) as directed by the Maxwell equation (8), we have

$$\left[E_y(x+\Delta x) - E_y(x)\right]\Delta y \simeq -\left.\frac{\partial B_z}{\partial t}\right|_{x+(\Delta x/2)} \Delta x \Delta y.$$

The left-hand side contains a term that represents the difference of a function of x taken at two closely spaced points, and we know from the meaning of a differential that it can be written as

$$\left.\frac{\partial E_y}{\partial x}\right|_x \Delta x,$$

which gives us

$$\left.\frac{\partial E_y}{\partial x}\right|_x \Delta x \Delta y \simeq -\frac{\partial B_z}{\partial t} \Delta x \Delta y. \qquad (12)$$

Taking the limit as the rectangle becomes small, we get an important equation:

$$\frac{\partial E_y(x,t)}{\partial x} = -\frac{\partial B_z(x,t)}{\partial t}. \qquad (13)$$

Now apply the same method to a similar rectangle in the xz-plane:

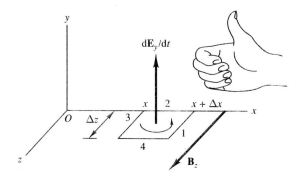

Diagram 4

We evaluate the left-hand side of the second Maxwell equation (9) as

$$\oint \mathbf{B} \cdot d\mathbf{s} = -B_z(x+\Delta x)\Delta z + 0 + B_z(x)\Delta z + 0.$$

Learning Guide 19   Maxwell's Equations and Electromagnetic Waves

As in (10), the nth term is the contribution of side n of the square. The right-hand side of (9) reduces to

$$\mu_0\epsilon_0 \iint \frac{\partial \mathbf{E}}{\partial t} \cdot d\mathbf{A} \simeq \mu_0\epsilon_0 \left.\frac{\partial E_y}{\partial t}\right|_{x+(\Delta x/2)} \Delta x \Delta z.$$

Combining them, we have

$$-[B_z(x+\Delta x) - B_z(x)]\Delta z \simeq \mu_0\epsilon_0 \left.\frac{\partial E_y}{\partial t}\right|_{x+(\Delta x/2)} \Delta x \Delta z.$$

Once again we recognize that the difference of B_z at two neighboring points can be written as

$$\left.\frac{\partial B_z}{\partial x}\right|_x \Delta x = B_z(x+\Delta x) - B_z(x).$$

So we get

$$-\left.\frac{\partial B_z}{\partial x}\right|_x \Delta x \Delta z \simeq \mu_0\epsilon_0 \frac{\partial E_y}{\partial t} \Delta x \Delta z, \tag{14}$$

or, in the limit, the second important equation:

$$-\frac{\partial B_z(x,t)}{\partial x} = \mu_0\epsilon_0 \frac{\partial E_y(x,t)}{\partial t}. \tag{15}$$

Suppose we take the derivative of (13) with respect to x and the derivative of (15) with respect to t. Then we get

$$\frac{\partial^2 E_y}{\partial x^2} = -\frac{\partial^2 B_z}{\partial x \partial t}; \quad -\frac{\partial^2 B_z}{\partial t \partial x} = \mu_0\epsilon_0 \frac{\partial^2 E_y}{\partial t^2}. \tag{16}$$

The two mixed derivatives that are just next to the semicolon are not necessarily equal. However, *in almost all physical applications*, functions like **B** and **E** are well-behaved and continuous, and under these circumstances it can be proved that *they are* equal. So

$$\frac{\partial^2 E_y(x,t)}{\partial x^2} = \mu_0\epsilon_0 \frac{\partial^2 E_y(x,t)}{\partial t^2}. \tag{17}$$

Eureka! We have produced a wave equation just exactly like the wave equation for transverse displacement waves on a stretched string,

$$\frac{\partial^2 y}{\partial x^2} = \frac{\mu}{\tau}\frac{\partial^2 y}{\partial t^2} = \frac{1}{v^2}\frac{\partial^2 y}{\partial t^2}, \tag{18}$$

for which we found that the wave speed v was equal to $\sqrt{\tau/\mu}$.

Both of these equations can be solved by functions of the form

$$y \text{ or } E = f(x - vt), \tag{19}$$

as you can see by direct substitution into (17) or (18). They can also be solved by the useful special functions

$$\cos\left[2\pi\left(\frac{x}{\lambda} \pm ft\right)\right]; \quad \sin\left[2\pi\left(\frac{x}{\lambda} \pm ft\right)\right], \qquad (20)$$

where

$$\lambda f = v.$$

In the case of electromagnetic waves,

$$v = \frac{1}{\sqrt{\mu_0 \epsilon_0}}. \qquad (21)$$

Let's look at this more carefully:

$$\epsilon_0 = 8.85418 \times 10^{-12} \text{ F/m}$$

$$\mu_0 = 4\pi \times 10^{-7} \text{ H/m}.$$

So, combining them with (21) and taking the square root gives

$$v = 2.997 \times 10^8 \sqrt{\frac{\text{m}^2}{\text{F} \cdot \text{H}}}$$

A farad is a coulomb per volt, and a henry is a volt per (coulomb per s^2), so the dimensions of v are, in fact, correct: m/s. Thus, we have found that the speed v is the speed of light, c, whose best experimental value is currently $2.997925 \pm 0.000003 \times 10^8$ m/s. (Actually, as you might guess, the real process is to measure c experimentally, and then use (21) to evaluate ϵ_0.)

Now what about the **B**-field? If we had differentiated (13) by t and (15) by x, we would have been able to combine them into a wave equation just like (17):

$$\frac{\partial^2 B_z(x,t)}{\partial x^2} = \mu_0 \epsilon_0 \frac{\partial^2 B_z(x,t)}{\partial t^2}. \qquad (22)$$

Thus **B** and **E** obey the same equation! Assume that we have a wave given by

$$E_y = E_0 \sin\left[2\pi\left(\frac{x}{\lambda} - ft\right)\right]. \qquad (23)$$

Then we can make use of (15), which is

$$\frac{\partial B_z}{\partial x} = -\mu_0 \epsilon_0 \frac{\partial E_y}{\partial t}. \qquad (15)$$

The right-hand side is

$$-\mu_0 \epsilon_0 (-2\pi f) E_0 \cos\left[2\pi\left(\frac{x}{\lambda} - ft\right)\right],$$

and if we integrate this with respect to x as implied by the left-hand side of (15), we find

$$B_z = \mu_0 \epsilon_0 f \lambda E_0 \sin\left[2\pi\left(\frac{x}{\lambda} - ft\right)\right] + \text{constant}$$

$$= B_0 \sin\left[2\pi\left(\frac{x}{\lambda} - ft\right)\right] + \text{constant}.$$

From the foregoing you can see that

$$B_0 = \frac{1}{c} E_0 \quad \text{since} \quad \mu_0 \epsilon_0 = \frac{1}{c^2}; \quad f\lambda = c. \tag{24}$$

B and **E** are evidently in phase with one another and can be represented by the next drawing.

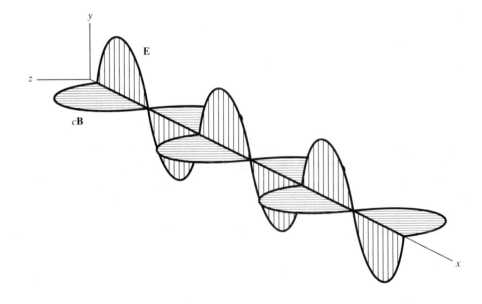

Diagram 5

Apparently, the propagation is in a direction found by taking **E** × **B** as positive.
A few things must yet be said about this picture:

1. This represents a monochromatic plane polarized wave. About the only place you are likely to find a real example of this kind of light is in a continuous gas laser beam or a microwave radar beam.
2. Ordinary white light, and even usual colored light, is an incoherent superposition of very many different wavelengths and all different directions of polarization. The difference between the former and the latter is that the tip end of the **E**-vector in diagram 5 would not appear to be sinusoidally oscillating in the yz-plane, but would look like the next figure. The laser oscillation shown on the right might be consid-

ered to be a single mode, and the white light on the left is a superposition of many modes.

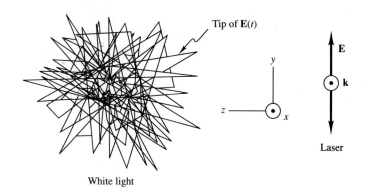

White light

3. In order to find plane waves, you must move relatively far away from the source as indicated by the next diagram, since the waves are radiated spherically outward from most sources.

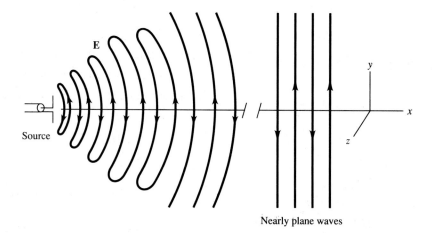

Nearly plane waves

ANSWER KEY

1. $f_c \simeq f(1 + v/c)$

2. $E = v/x$; $\Phi_E = EA$

3. πa^2

4. $F > 0$;

   $\dfrac{a^2}{m} > \dfrac{4cGM_\odot}{P_0}$.

5. $|S(x, t)| =$

   $\dfrac{E_0^2}{\mu_0 c} \cos^2(kx - \omega t + \phi)$

   $\langle S \rangle = \dfrac{1}{2} \dfrac{E_0^2}{\mu_0 c}$

6. $V'(t) = V(t)R^2/r^2 = (0.24 \text{ mV}) \sin(\omega t)$

7. $F_G = GM_\odot m/r^2$, toward the sun; where $M_\odot$ = sun's mass
8. $\frac{1}{2}$
9. Energy flux $= 2\pi al S = IV$
10. $B_\phi = \dfrac{\mu_0}{2\pi}\dfrac{r}{R^2} I_d$
    $= \dfrac{r\omega V_m}{2c^2 x}\cos(\omega t)$
    $= (2.5 \times 10^{-10}\text{ T})\cos(\omega t)$
11. $\Delta p = U/c$
12. $V(t)$:

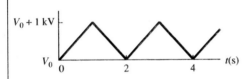

    Slope $= \pm 1$ kV/s.
13. $V(t) = N\displaystyle\oint_{\text{one turn}} \mathbf{E}\cdot d\mathbf{l} = -N\dfrac{d\Phi_B}{dt}$
    $= -N\dfrac{\pi d^2}{4}\dfrac{dB_\phi}{dt}$
14. $\mathbf{B} = B\mathbf{k}$, $\mathbf{B} = B(-\mathbf{j})$
15. $C = \epsilon_0 A/x$
16. $\langle S \rangle = 0.40\ \mu\text{W/m}^2$,
    $E_0 = 1.7 \times 10^{-2}$ V/m,
    $B_0 = E_0/c = 5.7 \times 10^{-11}$ T.
17. Faraday's law
18. $f_r \simeq f_c(1 + v/c)$
19. $S_{\text{wire}} = 0$
20. $\oint \mathbf{B}\cdot d\mathbf{l} = \displaystyle\int_0^{2\pi} B_\phi r\, d\phi =$
    $2\pi r B_\phi$
21. 7.3 Hz per mi/h
22. $\dfrac{dU}{dt} = \langle S(r)\rangle \pi a^2 = \dfrac{P_0}{4}\dfrac{a^2}{r^2}$
23. $\lambda = 3$ m;
    $V_m \simeq 3 \times 10^{-2}$ V.
    Yes, typically $V_{\text{noise}} \simeq 1\ \mu\text{V}$.
24. $V(t) = N\dfrac{\pi d^2}{4}\dfrac{r\omega^2 V_m}{2c^2 x}\sin(\omega t)$
    $= (0.2\text{ mV})\sin(\omega t)$
25. $I_d(t) = \epsilon_0 \dfrac{\pi R^2}{x}\omega V_m \cos(\omega t)$
    $= (0.53\text{ mA})\cos(\omega t)$
26. $dV/dt = 10^3$ V/s;
    $V = V_0 + (10^3\text{ V/s})\, t$;
    V_0 = arbitrary constant.
27. $\langle S(r)\rangle \pi a^2$
28. $I_1 = I_0/2$
29. $F_{\text{rad}} = \dfrac{dp}{dt} = \dfrac{1}{c}\dfrac{dU}{dt}$
30. Transmits it; absorbs it
31. $f_d = f_r = f_c\left(1 + \dfrac{v}{c}\right)$
    $= f\left(1 + \dfrac{v}{c}\right)^2$
    $\simeq f\left(1 + \dfrac{2v}{c}\right)$
32. $I_d(r) = I_d r^2/R^2$
33. $B = \mu_0 I/2\pi a$, around resistor;
    $E = \rho I/\pi a^2$, along resistor
34. $E_2 = E_1\cos\phi$; $I_2 = I_1\cos^2\phi$
35. $\langle S(r)\rangle = P_0/4\pi r^2$
36. The intensity out goes to zero.
37. $\dfrac{dp}{dt} = \dfrac{1}{c}\dfrac{dU}{dt}$
38. No. If you rotate polarizer 1, I does not change.
39. $E_3 = E_2\sin\phi$; $I_3 = I_2\sin^2\phi$
40. $F_{\text{rad}} = \dfrac{1}{c}\langle S(r)\rangle \pi a^2 = \dfrac{1}{c}\dfrac{P_0}{4}\dfrac{a^2}{r^2}$,
    outward.
41. $S = \dfrac{IV}{2\pi al} = \dfrac{I^2\rho}{2\pi^2 a^3}$,
    radially into the resistor.
42. $I_3 = \dfrac{I_0}{2}\cos^2\phi \sin^2\phi$
43. $B = \mu_0 I/2\pi r_0$ (Ampère's law);
    $E = 0$ (along a perfect conductor).
44. No. $a^2/m = 3.9$ m^2/kg, but $4cGM_\odot/P_0 = 410$ m^2/kg.
45. $I_2 = (I_0/2)\cos^2\phi$

46. The energy dissipated in the resistor can be thought of as entering through the **E**- and **B**-fields, not through the wire.

47. On the **E**-field

48. Joule heating $= IV$.

49. $F = -\dfrac{GM_\odot m}{r^2} + \dfrac{1}{c}\dfrac{P_0}{4}\dfrac{a^2}{r^2},$

outward.

learning guide 20

Light and Optics

Suggested Reading: Fishbane, Gasiorowicz, & Thornton, Chapters 36 and 37

PROBLEM I

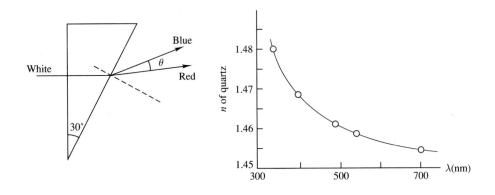

A collimated beam of white light is incident on a prism of fused quartz as shown. A plot of the index of refraction for quartz versus wavelength is also shown. Calculate the angle between emerging beams of red (4.6×10^{14} Hz) and blue (6.3×10^{14} Hz) light. You may need to look at Helping Questions 1 and 2.

Key 12

PROBLEM II

Two thin lenses of focal lengths f_1 and f_2 are in contact. Show that they are equivalent to a single thin lens of focal length f. Find f. See Helping Question 3 if you really need to.
Key 20

PROBLEM III

A penny lies at the bottom of a swimming pool of depth h. When viewed from directly above, how far below the surface does the penny appear? The index of refraction of water is $n = 1.33$.
Key 1

If help is needed, use Helping Questions 4 to 6 sparingly. (You should also reread Fishbane, Gasiorowicz, & Thornton, Section 36-3.)

PROBLEM IV

Verify, by explicitly tracing the rays (as in Fishbane, Gasiorowicz, & Thornton, Section 37-2) that the formula

$$\frac{1}{s} + \frac{1}{i} = \frac{2}{r}$$

holds for the case of a virtual image formed by reflection for a convex spherical mirror. Notice that r is taken as negative for convex mirrors and that i is taken as negative for virtual images.

If you can't work it out, use as few of Helping Questions 7 through 10 as you can.

PROBLEM V

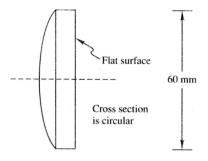

An actual lens made of crown optical glass is ground to the dimensions shown in the figure. Find the focal length. If you can't do it, Helping Questions 11 and 12 should help.
Key 9

PROBLEM VI

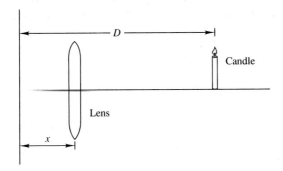

A candle is a distance D from a screen. A converging lens of focal length f is placed in between; $f < D/4$.

1. There are two positions of the lens for which a real image will be formed on the screen. Where are they? **Key 16**
2. What is the ratio of the image sizes for these two lens positions? **Key 26**

PROBLEM VII

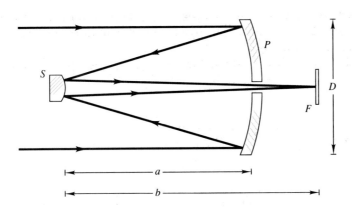

The optics for the Space Telescope are shown in the figure. Parallel light enters from the left, reflects off the primary mirror P and the secondary mirror S, and finally comes to a focus at F behind the primary mirror, where it is recorded by electronic cameras or analyzed by spectrographs.

The following information is derived from the specification given to the company building the space telescope optics:

Primary mirror diameter: $D = 2.40$ m;
Primary mirror focal length: $f_p = 5.51$ m;
Primary mirror–secondary mirror separation: $a = 4.90$ m;
Secondary mirror–focus separation: $b = 6.40$ m.

1. Where is the image formed by the primary mirror (ignore the secondary mirror for this part of the problem)? **Key 2**
2. What is the focal length of the secondary mirror? **Key 24**

(If you can't get it, see Helping Questions 13, 14, and 15.)

3. Two stars are separated by 10^{-4} rad. How far apart will their images be? (See Helping Question 16.) **Key 8**

PROBLEM VIII: THE EYE LENS AND THE MICROSCOPE

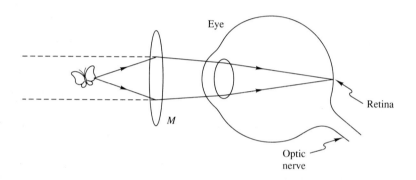

The Eye Lens (See also Fishbane, Gasiorowicz, & Thornton, Section 37-5)

The eye normally receives light rays from distant objects and focuses them to form an image on the retina. The light rays reaching the eye from a particular point on a distant object are nearly parallel, so the eye must focus all rays arriving at a given angle onto a given point on the retina. This is accomplished when the retina is one focal length behind the lens of the eye.

In order to magnify an object, it may be brought very close to the eye. However, most people are unable to see an object comfortably and sharply if it is brought closer than about 25 cm, which is arbitrarily decreed to be a "comfortable" viewing distance.

1. If a lens M of focal length 2.5 cm is used (see the figure) as a magnifying lens placed close to the eye (< 2.5 cm away), what is the magnification m? That is, how much larger will the object appear when held at a focal point of M than when held at the normal viewing distance of 25 cm? **Key 31**

If necessary, use Helping Questions 17 and 18.

Such a magnifying glass is called an *eye lens* (or *eyepiece*, or *ocular*) when used as part of various optical instruments.

Learning Guide 20   Light and Optics

The Microscope (See also Fishbane, Gasiorowicz, & Thornton, Section 37-5)

Consider a brightly illuminated small object at *A* and a translucent screen, *S*, 22.5 cm away from it.

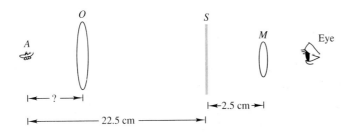

2. Where must lens *O* with focal length of 16 mm be placed in order that the magnified image of *A* be located exactly on *S*? Helping Question 19 may direct you. **Key 10**

3. Now suppose that lens *M* ($f = 25$ mm) is used as an eye lens to examine the image of *A* on the screen *S*. What is the total magnification of the entire combination? If necessary, use Helping Question 20. **Key 17**

4. If the screen *S* is removed, you have a compound microscope. What effect do you think removing the screen has? **Key 5**

5. Can a microscope have an arbitrarily large magnification? **Key 14**

HELPING QUESTIONS

1. What is the formula for angle of refraction in terms of the index of refraction and the angle of incidence? If your answer doesn't match, review Fishbane, Gasiorowicz, & Thornton, Section 36-3. **Key 18**

2. What wavelengths correspond to frequencies of 4.6×10^{14} Hz and 6.3×10^{14} Hz? **Key 25**

3. What are the object and image distances for the single lens? Then apply the thin-lens equations (Fishbane, Gasiorowicz, & Thornton, Eqs. (37-10) and (37-11)). **Key 28**

4. Trace two rays leaving the penny. Let one go straight up, and let the other make a small angle θ to the vertical. **Key 27**

5. How is θ' (the angle of the refracted ray) related to θ? **Key 19**

6. The two rays cross the surface of the water at some distance *a* apart. Express *a* in terms of *h* and θ, and in terms of h' and θ', where h' is the apparent depth. **Key 11**

7. Draw a diagram showing all the relevant points, angles, and lengths. (Do *not* look at the key until you have drawn and labeled your own diagram. Your labels will no doubt be different from the key, but the diagrams should be essentially identical.) **Key 7**

8. Derive an exact equation relating α, β and γ (see diagram of **Key 7**). If you're success-

ful at this, but you still can't finish the proof, skip to Helping Question 10. **Key 23**

9. The exterior angle of a triangle is equal to the sum of the two opposite interior angles. Apply this theorem to the triangles OAC and OAI (see diagram of **Key 7**). **Key 15**

Retry Helping Question 8.

10. Derive approximate or exact relations for α, β, and γ in terms of the lengths in the diagram. Use radian measure, of course. Satisfy yourself that the approximations are valid for paraxial rays, that is, when the angles are small. **Key 30**

11. What are the radii of curvature of the front and back surfaces? **Key 6**

12. Assuming these surfaces are close together, what does Fishbane, Gasiorowicz, & Thornton, Eq. (37-11), have to say?

13. Think of the image formed by the primary mirror as the source for the secondary. Is it a real or a virtual source? Why? **Key 29**

14. According to the convention, is the source distance to the secondary mirror positive or negative? **Key 21**

15. If you aren't comfortable with the concept of a "virtual source," reverse all of the arrows in the diagram and think of the point F as a light source. Now, to have parallel light after finally reflecting off mirror P, where must the virtual image of F formed by mirror S be? **Key 3**

16. First reason that the images of the two stars formed by mirror P will be separated by 5.51×10^{-4} m, and then multiply this by the magnification of mirror S. What is m? **Key 32**

17. If an object of size Y subtends an angle θ' at 25 cm, what angle θ will it subtend at 2.5 cm? **Key 22**

18. How is magnification of a magnifier defined? **Key 13**

19. What are s, i, and f? **Key 33**

20. How much larger is the image of A on S than A itself? **Key 4**

ANSWER KEY

1. $h' = h \dfrac{\tan \theta_2}{\tan \theta_1} \simeq \dfrac{h}{n} = \dfrac{h}{1.33}$

2. 5.51 m to the left of the primary mirror

3. At the focal point of P, that is, 0.61 m to the left of S

4. $OS/AO = 12$

5. The image seems brighter and clearer, but the field of view is smaller.

6. 8 cm and ∞

7.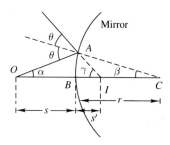

8. 5.8 mm

9. 15 cm, or 6 in

10. $\frac{1}{x} + \frac{1}{(22.5-x)} = \frac{1}{1.6}$;
    $x = 1.73$ cm
11. $a = h\tan\theta \simeq h\theta$
    $a = h'\tan\theta' \simeq h'\theta'$
12. $\theta_b - \theta_r = \sin^{-1}\left(\frac{1.463}{2}\right) - \sin^{-1}\left(\frac{1.457}{2}\right)$
    $\simeq 0.25° = 0.004$ rad
13. $m = \theta'/\theta$
14. No, diffraction will eventually blur the image.
15. $\theta = \alpha + \beta$
    $2\theta = \alpha + \gamma$
16. $x = \frac{D}{2}\left(1 \pm \sqrt{1 - 4\frac{f}{D}}\right)$
17. $OS/AO \times 10 = 12 \times 10 = 120$
18. $n\lambda \sin 30° = \sin[\theta(\lambda)]$
19. $n\sin\theta = \sin\theta' \Rightarrow n\theta \simeq \theta'$
20. $\frac{1}{f} = \frac{1}{f_1} + \frac{1}{f_2}$
21. Negative—it is virtual.
22. $\Theta' = \tan^{-1}\left(\frac{y}{2.5}\right) \simeq y/2.5 = 10\theta$ for small angles.
23. $\alpha + 2\beta = \gamma$
24. -0.67 m
25. $\lambda = c/f$:

4.6×10^{14} Hz $\to 650$ nm;

6.3×10^{14} Hz $\to 480$ nm

26. $\dfrac{\left[(D/2) + \sqrt{(D/2)^2 - fD}\right]^2}{\left[(D/2) - \sqrt{(D/2)^2 - fD}\right]}$

27.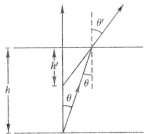

28. $s = s_1$; $i = i_2$
29. Virtual, because no light actually emanates from the source point, since it strikes the secondary mirror before it can fully converge.
30. $\alpha = \dfrac{AB}{s}$, $\beta = -\dfrac{AB}{r}$, $\gamma = \dfrac{AB}{i}$
31. $25/2.5 = 10$
32. $m = -\dfrac{i}{s} = \dfrac{-6.40 \text{ m}}{-0.61 \text{ m}} = 10.5$
33. Use $s = AO$, $i = OS$, $f = 1.6$ cm.

learning guide 21

Interference and Diffraction

Suggested Reading: Fishbane, Gasiorowicz, & Thornton, Chapters 38 and 39

PROBLEM I

A double-slit arrangement produces interference fringes for sodium light (wavelength in air $\lambda = 589$ nm) that are $0.30°$ apart. What is the angular fringe separation if the entire arrangement is immersed in water ($n = 1.33$ in water)?

Key 9

If you are correct, proceed to Problem II. If, however, you are still confused after an honest effort, try Helping Questions 1 and 2. If you remain lost after those hints, try Helping Questions 3 and 4.

PROBLEM II

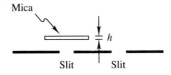

A thin flake of mica ($n = 1.58$) is used to cover one slit of a double-slit arrangement. The central point of the viewing screen is occupied by what used to

be the sixth bright fringe. If $\lambda = 550$ nm, what is the thickness h of the mica?
Key 17

If you encounter difficulties, try Helping Questions 5 through 9. Also, review Fishbane, Gasiorowicz, & Thornton, Sections 38-1 and 38-2.

PROBLEM III

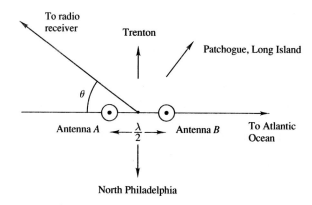

A radio station situated between Trenton and North Philadelphia wants to broadcast into these two areas and to beam as little power toward the ocean (due east) as possible. Its two antennas are separated by $\lambda/2$ and the radio waves are sent out in phase. The antennas radiate at equal amplitude in all directions in the plane of the earth.

1. Compute the intensity that a radio at a given distance would receive as a function of θ. Assume that the radio is much farther away than the antenna separation. **Key 10**

   If you have difficulty see Helping Questions 10 through 12.

2. What is the intensity in the direction of Patchogue ($\theta = 120°$) relative to that toward Trenton? **Key 16**
3. What is the intensity toward the Atlantic Ocean ($\theta = 180°$)? **Key 7**
4. The power toward Trenton is 4 times that of a single antenna. Doesn't this violate energy conservation? **Key 20**

PROBLEM IV

In a single-slit diffraction pattern the distance between the first minimum on the right and the first minimum on the left is 5.6 mm. The screen on which the pattern is displayed is 80 cm from the slit, and the wavelength is 5460 Å (1 Å = 0.1 nm = 10^{-10} m). Calculate the slit width. **Key 1**

If your answer is incorrect, see Helping Questions 13 and 14.

Learning Guide 21   Interference and Diffraction

PROBLEM V

You are in charge of the satellite spy branch of the CIA. The DCI (Director of Central Intelligence) asks you: From a satellite at 100 km, how large a telescope mirror must you put in orbit to be able to tell

1. The brand of cigarettes that the bad guys are smoking? **Key 6**
2. Whether they are carrying pistols or submachine guns? **Key 2**
3. Whether an object is a missile or a tanker truck? **Key 4**

If you're puzzled, try Helping Questions 15 and 16.

PROBLEM VI

A grating is ruled with 400 lines (slits) per millimeter and is set up in vacuum to examine a monochromatic source of ultraviolet light. The first-order diffraction is observed at 4.5° from the central maximum. To what wavelength λ does this correspond? **Key 33**

If you have difficulty, try Helping Questions 17 and 18.

PROBLEM VII

A four-slit grating has separation d between adjacent slits. If one of the outer slits is covered up, will the half width of the intensity maxima become broader or narrower? **Key 12**

If you are wrong or unsure, review Fishbane, Gasiorowicz, & Thornton, Section 39-2.

PROBLEM VIII

A source containing a mixture of hydrogen and deuterium atoms emits a red doublet at $\lambda = 6563$ Å whose separation is 1.8 Å. Find the minimum number of rulings needed in a diffraction grating to resolve these in the first order. **Key 39**

Reread Fishbane, Gasiorowicz, & Thornton, Section 39-2, if you need help.

PROBLEM IX

Light of wavelength $\lambda = 600$ nm is incident normally on a diffraction grating. Two *adjacent* principal maxima occur at $\sin\theta = 0.10$ and $\sin\theta = 0.20$, respectively. The fourth order is a missing order.

1. What is the separation d between adjacent slits? **Key 28**

If you encounter any difficulty, try Helping Questions 19 and 20.

2. What is the smallest possible individual slit width a? **Key 24**

For assistance, see Helping Questions 21 through 23.

3. Name all orders that actually appear on the screen with the values chosen in parts (1) and (2). **Key 34**

If your answer doesn't check, use Helping Question 24.

HELPING QUESTIONS

1. How is λ', the wavelength of sodium light in water, related to λ, its wavelength in air? **Key 5**

If you don't see this, review Fishbane, Gasiorowicz, & Thornton, Section 38-3.

2. For small θ, how are $\sin\theta$ and θ related? **Key 19**

Retry Problem I.

3. Let d be the separation between the slits. How is $\Delta\theta$, the angular fringe separation in air, related to d and λ? **Key 3**

If your answer is incorrect, review Fishbane, Gasiorowicz, & Thornton, Section 38-1.

4. How is $\Delta\theta'$, the angular fringe separation in water, related to d and λ'? **Key 18**

Retry Problem I.

5. What does the mica do to the wavelength? **Key 11**

If you don't get this, look at Eq. (36-1) in Fishbane, Gasiorowicz, & Thornton.

6. What is the phase difference for the sixth bright fringe? **Key 21**

7. How many wavelengths are there in a "slab" of air that has the same thickness h as the mica? **Key 8**

8. How many wavelengths are there in a slab of mica of thickness h? **Key 15**

9. What is the difference in the phase with and without the mica is what? **Key 23**

10. How is the phase difference ϕ between the radio waves from the two antennas related to the antenna separation and the direction of the receiver? **Key 14**

11. The extra distance that the wave from antenna B must travel is $(\lambda/2)\sin\theta$. Right? **Key 30**

12. The amplitude of the total wave is $A = 2A_0\cos(\phi/2)$, where ϕ is the phase difference and A_0 is the single-wave *amplitude*. What is the average *intensity*? **Key 13**

13. What is the condition for the first minimum? If you are confused, review Fishbane, Gasiorowicz, & Thornton, Section 39-3. **Key 25**

14. What is $\sin\theta$ in terms of the given quantities? **Key 32**

15. What size must you be able to resolve in each case? **Key 38**

16. What is the relation between the critical angular separation α_c, the wavelength λ, and the diameter D of the circular opening (here, the mirror)? If you are unsure, read Fishbane, Gasiorowicz, & Thornton, Section 39-4, again. **Key 27**

17. The first-order maximum occurs for a path-length difference of λ between the paths from adjacent slits. Right? **Key 35**

18. What is the path-length difference between adjacent slits? **Key 26**

19. If one maximum is order m and the next is order $m+1$, write down two equations for d. **Key 22**

20. Now eliminate m from these equations by subtracting them.

21. At what value of $\sin\theta$ should the fourth order occur? **Key 36**

22. What value of m in Fishbane, Gasiorowicz, & Thornton, Eq. (39-11), would make the slit width smallest? **Key 29**

23. What is the slit width, then, in terms of the answers to Helping Questions 21 and 22? **Key 37**

24. With a and λ now known, does Eq. (39-11) predict any other missing orders beside those using the value of m in Helping Question 22? If so, for what value of $\sin\theta$? **Key 31**

ANSWER KEY

1. $a = \dfrac{\lambda}{\sin\theta} = \dfrac{0.8\lambda}{2.8\times 10^{-3}}$
   $= 0.156$ mm

2. 0.5 m in diameter, a reasonable size to build. The Princeton telescope is 1 m in diameter.

3. $d\Delta\theta = \gamma$

4. 5 cm in diameter. This is easy to build.

5. $\lambda' = \lambda/n$

6. About 22 m in diameter, which is impossible to build. The largest telescope on earth is 6 m in diameter.

7. 0

8. h/λ

9. $\Delta\theta' = 0.30/n \simeq 0.23°$

10. $I = 2I_0[1+\cos(\pi\cos\theta)]$, where I_0 is the single-wave intensity at the given distance.

11. Wavelength decreases in mica. λ_{vacuum} is replaced by $\lambda_{mica} = \lambda_{vacuum}/n$.

12. Broader

13. Intensity $= \tfrac{1}{2}|\text{amplitude}|^2$
    $= 2A_0^2\cos^2\left(\tfrac{1}{2}\pi\cos\theta\right)$
    $= A_0^2[1+\cos(\pi\cos\theta)]$

14. $\Phi = \dfrac{2\pi}{\lambda}\left(\dfrac{\lambda}{2}\cos\theta\right) = \pi\cos\theta$

15. nh/λ

16. $I_{\text{Trenton}}/I_{\text{Patchogue}} = 2$

17. $h = 6\lambda/(n-1) = 5.69\times 10^{-6}$ m $= 5.69\ \mu$m

18. $d\Delta\theta' = \lambda'$

19. $\sin\theta \simeq \theta$

20. No. Power in other directions is less than before (e.g., toward the Atlantic Ocean).

21. $6\times 2\pi$

22. $d(0.10) = m(600$ nm$)$,
    $d(0.20) = (m+1)(600$ nm$)$

23. $2\pi h(n-1)/\lambda$

24. $a = 1.5\times 10^3$ nm

25. $\lambda = a\sin\theta$

26. $d\sin\theta$, where d is the spacing of the slits.

27. $\alpha_c = 1.22\lambda/D$

28. $d = 6\ \mu$m

29. $m = 1$

30. Wrong, it is $(\lambda/2)\cos\theta$.

31. Yes; for example, $m = 2$ gives $\sin\theta = 0.8$, the eighth-order spectrum of the grating.

32. $\sin\theta = \dfrac{5.6/2 \times 10^{-3}}{0.8}$
    $= 3.5 \times 10^{-3}$

33. $\lambda = 196.1$ nm $= 1961$ Å

34. 0, 1, 2, 3, 5, 6, 7, 9; the tenth order is at $\theta = 90°$.

35. Right

36. $\sin\theta = 4\lambda/d = 0.4$

37. $a = (1)(600 \text{ nm})/0.4$

38. (a) About 0.3 cm; (b) about 10 cm; (c) about 1 m

39. $\dfrac{\lambda}{\Delta\lambda} = Nm = \dfrac{6563}{1.8}$
    $= 3646$ lines
    ($m = 1$, first order).

learning guide 22

Quantum Physics

Suggested Reading: Fishbane, Gasiorowicz, & Thornton, Chapter 41.

PROBLEM I

Solar radiation falls on the earth at the rate of 1 kW/m^2. Assume that solar radiation is concentrated at a wavelength of 5500 Å.

1. How many photons fall on 1 m^2 of the earth's surface in 1 s? **Key 7**

   See Helping Question 1 if you have to.

2. The sun would still be visible to the naked eye if it were 1000 light-years away. If the dilated pupil of the human eye is taken to have an area of 0.25 cm^2, how many photons per second would enter the eye from such a star? (You need to know that the earth-sun distance is 8 light-minutes, that 1 year is 3×10^7 s, and that 1 light-year is the distance traveled by light in 1 year.) If need be, use Helping Questions 2 and 3. **Key 13**

PROBLEM II: A QUANTUM ROTATOR

A diatomic molecule can be modeled as a rigid dumbbell with moment of inertia I rotating about its center of mass at angular frequency ω.

1. What is the angular momentum of the molecule? **Key 3**
2. The rule to quantize this system is to allow only angular momenta L that satisfy the equation

$$L = \sqrt{J(J+1)}\,\frac{h}{2\pi}; \qquad J = 0, 1, 2 \ldots.$$

where J is the quantum number for angular momentum. What are the allowed *energies* of the molecule? If necessary, use Helping Questions 4 and 5. **Key 8**
3. The longest wavelength *strongly* absorbed by carbon monoxide (CO) is $\lambda = 0.26$ cm. Compute the distance between the two atoms in carbon monoxide. You may wish to consult Helping Questions 6 through 8. **Key 1**

PROBLEM III

Suppose that we want to remove an electron from a hydrogen atom via the photoelectric effect.

1. What is the minimum photon energy necessary? That is, what is the work function ϕ of hydrogen? If you need to, use Helping Question 9. **Key 16**
2. What is the cutoff frequency? What is the corresponding wavelength? **Key 11**

PROBLEM IV

Apply Bohr's theory to the positronium "atom." This consists of a positive and a negative electron revolving around their center of mass, which lies halfway between them.

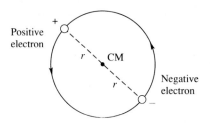

1. What is the force on either particle? What is the potential energy of the system? **Key 15**
2. What is the centripetal acceleration? **Key 6**
3. What is the velocity of either particle? If you need assistance, see Helping Question 10. **Key 18**
4. What is the kinetic energy of either particle? **Key 2**

5. What is the total energy E of the system? **Key 10**
6. What is the angular momentum of either particle? **Key 12**
7. What is the total angular momentum? How should it be quantized? (See Fishbane, Gasiorowicz, & Thornton, Section 41-4.) **Key 4**
8. What are r_n and E_n? How do these compare with hydrogen orbits and energies? **Key 14**
9. What is the relationship between this spectrum and the spectrum of hydrogen? **Key 17**

PROBLEM V

The de Broglie wavelength $\lambda = h/p$ of a neutron changes if its momentum p changes. If the neutron is moving in the gravitational field of the earth, its momentum, and therefore its wavelength, changes with altitude.

1. Suppose that a neutron is produced at altitude Z_0 with momentum p_0 and de Broglie wavelength $\lambda_0 = h/p$. Show that if it changes altitude by a small amount ΔZ, the de Broglie wavelength changes by

$$\Delta \lambda = \frac{\lambda_0 (mg\, \Delta Z)}{2(p_0^2/2m)}$$

or

$$\frac{\Delta \lambda}{\lambda_0} = \frac{1}{2} \frac{\Delta(\text{potential energy})}{\text{kinetic energy}}.$$

See Helping Questions 11 to 13 for assistance.
A way to exploit this wavelength change is in the following interferometer:

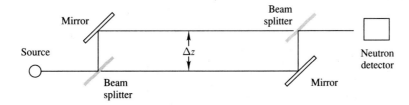

The two neutron paths to the detector are a different number of wavelengths long, and hence we should see interference effects at the detector.

2. By how much must we change L in order to make the detector signal go from a maximum to the nearest minimum, taking $\Delta Z = 1$ cm and letting the neutrons emerge from the source with energy 1 eV? (The mass of the neutron is 1.68×10^{-27} kg.) See Helping Questions 14 through 17 for some hints. **Key 24**

PROBLEM VI

An experimenter wants to do a double-slit interference experiment with a light source so feeble that there is only one photon at a time in his apparatus. He has a 1-W source of 5500-Å photons and a number of filters, each of which transmits 1% of the light incident on it. The apparatus looks like this:

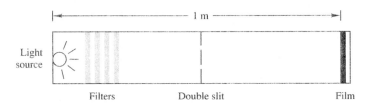

How many filters must he insert to achieve his goal of reducing the photon density to the point that, at any instant, there is on average only one photon in the apparatus? For assistance, try Helping Questions 18 through 21.
Key 28

Since on the average there is only one photon at a time in the apparatus, will he obtain an interference pattern on the film (if he allows enough time for many photons to strike the film)?
Key 22

HELPING QUESTIONS

1. If a photon has energy $hf = hc/\lambda$, what is the energy (in joules) of a photon of wavelength 5500 Å? (1 Å $= 10^{-1}$ nm $= 10^{-10}$ m.) **Key 20**

2. If a source at distance R produces an energy flux I, the same source at distance R' produces a flux $I' = I(R/R')^2$. Right? **Key 32**

3. What is the ratio (R/R') appropriate to comparing the fluxes produced by the sun at 8 light-minutes and 1000 light-years? **Key 35**

4. Write the energy as a function of ω. Then eliminate ω and find energy as a function of angular momentum L.

5. Substitute in the quantized L values.

6. What are the allowed energies for photon absorption? **Key 9**

7. Compute the energies of transition from E_0 to E_1, E_1 to E_2, E_2 to E_3, and so forth. Do you see a pattern? Which is the *smallest*? **Key 19**

8. The wavelength gives the energy for the $0 \to 1$ transition. Compute I. Relate this to the bond length of CO, considered as two point atoms of masses 12 and 16.

9. Consider the electron to be in the ground state ($n = 1$). What is the energy of this state? **Key 5**

10. $F = ma$ implies
$$\frac{e^2}{4\epsilon_0 4 r^2} = \frac{mv^2}{r}.$$

11. What is $\Delta(p^2/2m)$? **Key 31**

12. What is $\Delta\lambda$? **Key 34**

Learning Guide 22 Quantum Physics

13. Combining the last two hints yields
$$\frac{\Delta\lambda}{\lambda} = \frac{m^2 g \, \Delta Z}{p^2} = \frac{1}{2}\frac{(mg \, \Delta Z)}{(p^2/2m)}.$$

14. What is the de Broglie wavelength of a 1-eV neutron? **Key 27**

15. What is the phase difference between the upper and lower paths? **Key 33**

16. What is $\Delta\lambda/\lambda$ in the case at hand? **Key 29**

17. In order for the phase difference to increase by π, by how much must L increase? **Key 23**

18. How many photons per second are produced by the 1-W source? **Key 30**

19. How long does it take one photon to travel the 1 m from the light to the film? **Key 21**

20. Without filters, how many photons are in flight between bulb and film? **Key 26**

21. By what factor does one filter change the number of photons in flight? **Key 25**

ANSWER KEY

1. $r = 1.128$ Å (actually, $\lambda = 0.260078$ cm and $r = 1.128227$ Å).

2. $K = \frac{1}{2}mv^2 = \frac{e^2}{32\pi\epsilon_0 r}$

3. Angular momentum $L = I\omega$.

4. $L_{\text{total}} = \sqrt{\frac{re^2 m}{4\pi\epsilon_0}} = n\hbar$; $\left(\hbar = \frac{h}{2\pi}\right).$

5. -13.6 eV

6. $a = \frac{1}{4\pi\epsilon_0}\frac{e^2}{m(2r)^2}$

7. 2.8×10^{21} photons/m²·s

8. $E_J = \frac{1}{2I}\left(\frac{h}{2\pi}\right)^2 J(J+1)$; $J = 0, 1, 2 \ldots$

9. They must be those of a transition from one state to another: $E_J - E_{J'} = hf$ (second Bohr postulate).

10. $E = 2K + U = -e^2/16\pi\epsilon_0 r$

11. $f_t = \phi/h = 3.3 \times 10^{15}$ Hz; $\lambda_t = c/f_t = 920$ Å

12. $L = mvr = \frac{1}{2}\sqrt{\frac{re^2 m}{4\pi\epsilon_0}}$

13. About 18 photons per second!

14. $r_n = n^2\left(\frac{h^2\epsilon_0}{\pi e^2 m}\right) = r_n$ for H; $E_n = \frac{1}{n^2}\left(\frac{-me^4}{16\epsilon_0^2 h^2}\right) = \frac{1}{2}E_n$ for H.

15. $F = \frac{e^2}{4\pi\epsilon_0(2r)^2}$; $U = \frac{-e^2}{4\pi\epsilon_0(2r)}$

16. 13.6 eV $= \frac{me^4}{8\epsilon_0^2 h^2}$

17. $f = \frac{me^4}{16\epsilon_0^2 h^3}\left[\frac{1}{n_2^2} - \frac{1}{n_1^2}\right] = \frac{1}{2}f_H$

18. $v = \sqrt{\frac{e^2}{16\pi\epsilon_0 rm}}$

19. E_0 to E_1 is smallest energy change and hence the longest wavelength.

20. $hf = 3.6 \times 10^{-19}$ J

21. 0.3×10^{-8} s
22. Yes! (Principle of superposition.)
23. $\Delta L = (\lambda/2)(\lambda/\Delta\lambda)$
    $= \dfrac{1}{2} \dfrac{2.9 \times 10^{-11}}{5.1 \times 10^{-10}} = 2.85$ cm
24. $\Delta L \simeq 3$ cm
25. 10^{-2}
26. About 10^{10} photons
27. 2.9×10^{-11} m
28. Five filters
29. $\dfrac{\Delta\lambda}{\lambda}$
    $= \dfrac{1}{2} \dfrac{1.68 \times 10^{-27} \times 9.8 \times 10^{-2}}{1.6 \times 10^{-19}}$
    $= 5.1 \times 10^{-10}$
30. 2.8×10^{18} photons per second
31. $\Delta\left(\dfrac{p^2}{2m}\right) = -mg\,\Delta Z$
    or $p\,\Delta p = -m^2 g\,\Delta Z$.
32. Right
33. $\delta = 2\pi\left(\dfrac{L}{\lambda} - \dfrac{L}{\lambda + \Delta\lambda}\right)$
    $\simeq 2\pi\left(\dfrac{L}{\lambda}\right)\left(\dfrac{\Delta\lambda}{\lambda}\right)$
34. $\Delta\lambda = \Delta\left(\dfrac{h}{p}\right)$
    $= -h\Delta p/p^2 = -\lambda\Delta p/p$
    or $\dfrac{\Delta\lambda}{\lambda} = -\dfrac{\Delta p}{p}$.
35. $\dfrac{R}{r'} = \dfrac{8 \text{ light-minutes}}{1000 \text{ light-years}}$
    $= \dfrac{8 \times 60}{1000 \times 3 \times 10^7}$
    $= 1.6 \times 10^{-8}$